Grade 4

Mathematics

Pupil Book 4A

PAT LILBURN
PAM RAWSON
PETER SULLIVAN

in consultation with
Elsie Kinavai
Alex Feeger

Department of Education Papua New Guinea

253 Normanby Road, South Melbourne, Victoria 3205, Australia

Oxford University Press is a department of the University of Oxford. It furthers the University's objective of excellence in research, scholarship, and education by publishing worldwide in

Oxford New York

Auckland Cape Town Dar es Salaam Hong Kong Karachi Kuala Lumpur Madrid Melbourne Mexico City Nairobi New Delhi Shanghai Taipei Toronto

With offices in

Argentina Austria Brazil Chile Czech Republic France Greece Guatemala Hungary Italy Japan Poland Portugal Singapore South Korea Switzerland Thailand Turkey Ukraine Vietnam

OXFORD is a trademark of Oxford University Press in the UK and in certain other countries

First published 1999
Reprinted 2003, 2008 (twice), 2010

ISBN 978 0 19 550862 8

Written by Pat Lilburn, Pam Rawson and Peter Sullivan
Writing consultants: Elsie Kinavai and Alex Feeger of the Curriculum Development Division of the Papua New Guinea Department of Education
Editorial and design consultant: John Hughes of the Papua New Guinea Department of Education
Cover and text design by Mary Kerr
Cover artwork by Gigs Wena
Illustrated by Petra Hanzak, Boris Silvestri and Wendy Gorton
Printed in China by Golden Cup Printing Co. Ltd
Published by Oxford University Press
Editorial Office: PO Box 7979, Boroko NCD,
Papua New Guinea

CONTENTS

SECRETARY'S MESSAGE

This pupil book is part of the new reformed mathematics program designed and written for use in Grade 4 classes in Community and Primary schools throughout Papua New Guinea. The core materials consist of two pupil books called *Grade 4 Mathematics Pupil Book* 4A and *Grade 4 Mathematics Pupil Book* 4B. They are accompanied by the *Grade 4 Mathematics Teacher's Resource Book*. They replace the MACS series now in use.

In the Mathematics program children are first taught mathematics by using real objects. Later, the children will use pictures of objects. Finally, the children will use number symbols to represent these objects. Always allow your pupils to use real objects during their mathematics lessons if they want to. Teach the concepts in the context of real life situations as this leads to an appreciation of the everyday use of mathematics skills and knowledge. The children will decide when they do not need the help of real objects any longer. Remember, when children use real objects, they will understand mathematics better. This program also encourages the children to solve their own problems and make their own decisions with confidence. In order to learn these skills, the children should talk about what they are doing in every lesson. Since learning is most effective when it has meaning and is enjoyable, allow the children to use the language that they are most comfortable with.

However, in Grade 4 more English should be used together with the vernacular and other languages to aid understanding of new and difficult concepts.

Finally, the National Department of Education wants teachers to be flexible in programming and timetabling to cater for the different abilities of children and multigrade teaching situations. This book shows you some strategies to help satisfy these needs.

Peter M. Baki

PETER M. BAKI
Secretary of Education

Write the answers to these questions in your exercise book.

1. Find the mistake.

2. Which one is missing?

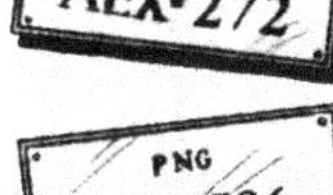

3. Write the numbers of the runners, in order.

4. What are the next four serial numbers?

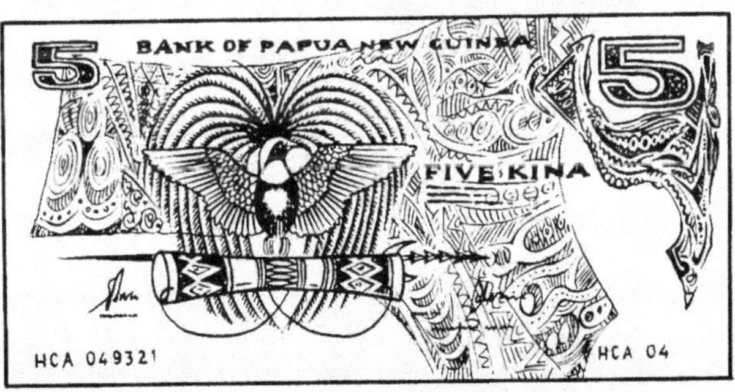

5. Think of some places where you see numbers.

What do these numbers tell us?

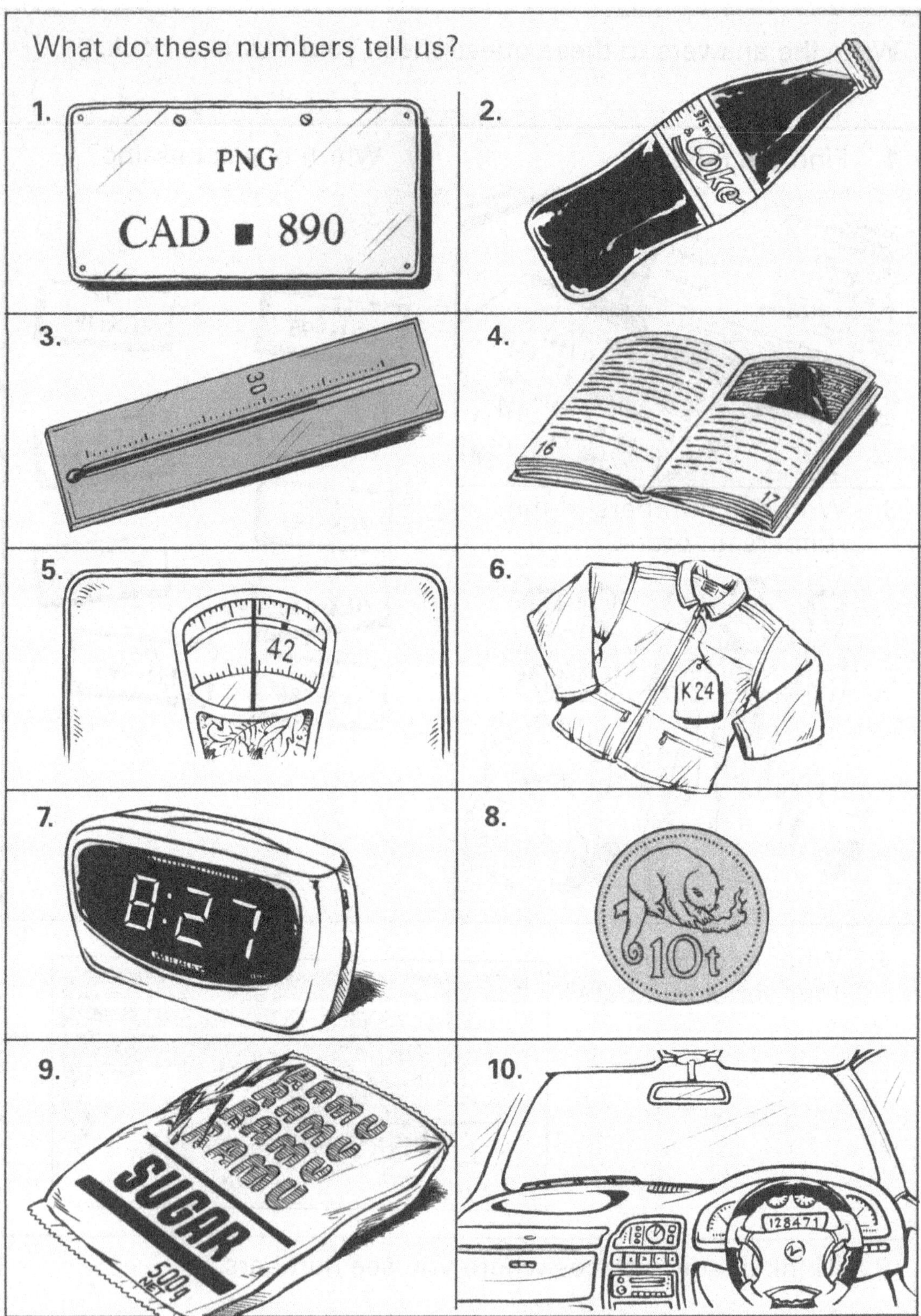

In your exercise book, write down the numbers and put them in order.

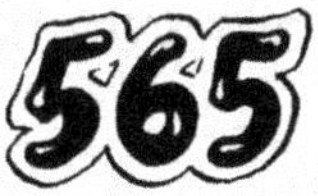

565

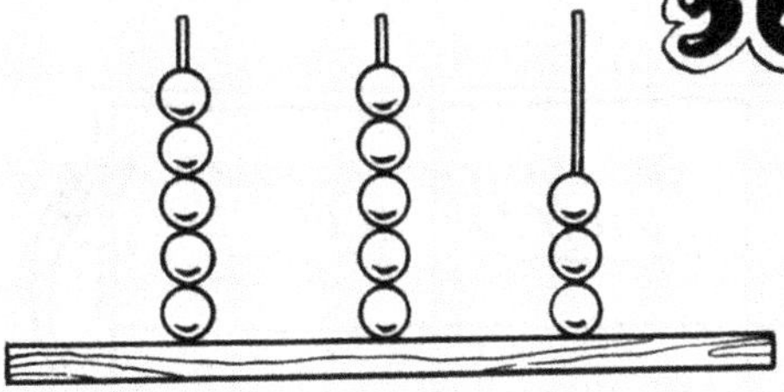

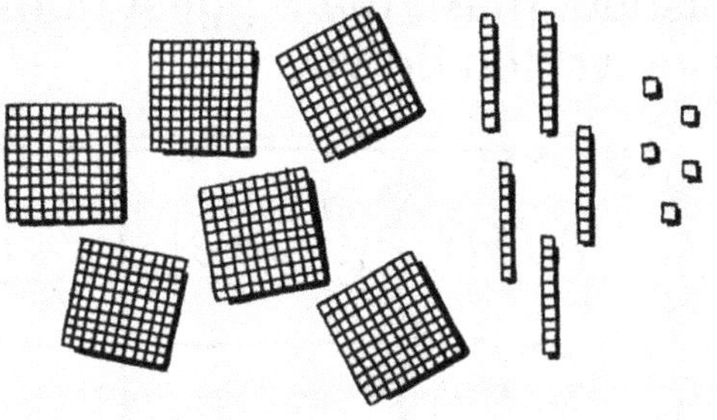

Six hundred and sixty five

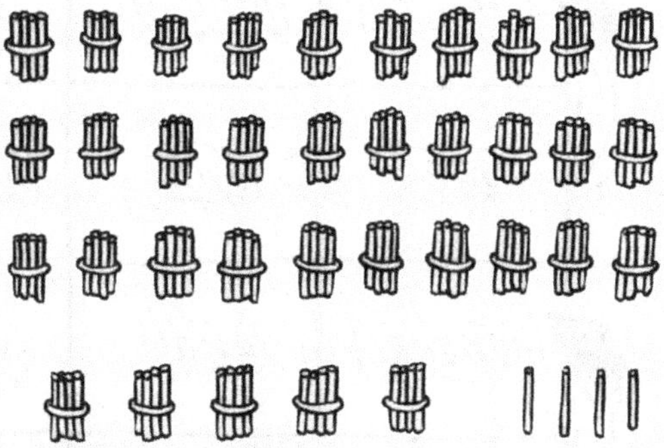

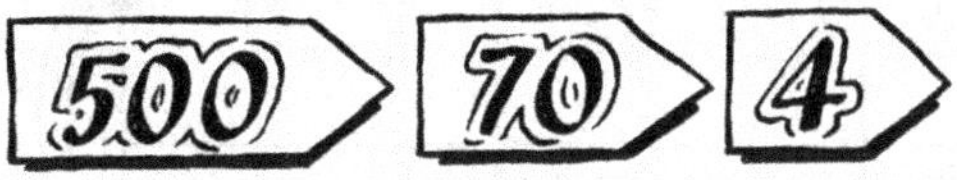

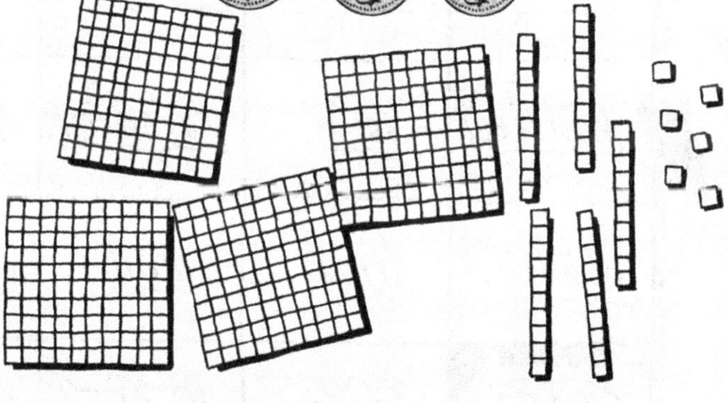

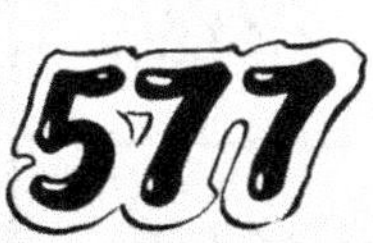

500 70 4

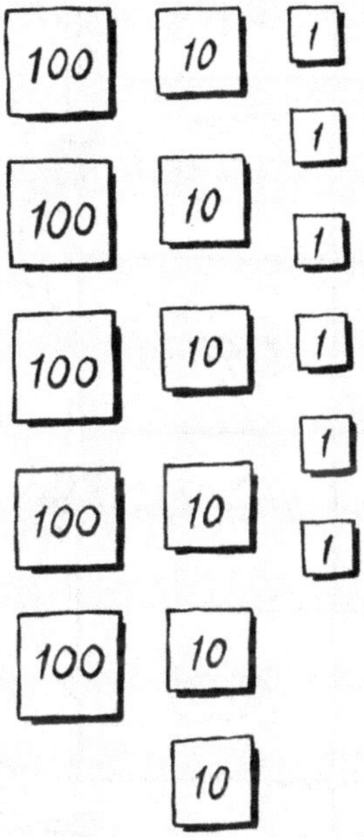

100	10	1
100	10	1
100	10	1
100	10	1
100	10	1
	10	1

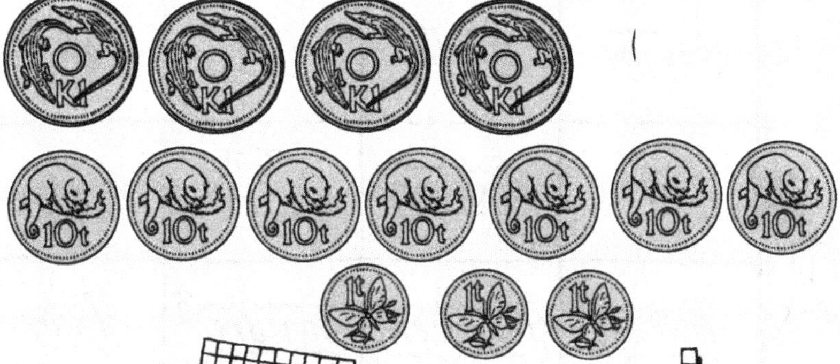

577

Write down the number, one more than each of these numbers shown.

Throw a dice. Move forward that number of squares. Each time, write down the word you land on. When you have finished, make the biggest number you can from the words you have written down.

FINISH	twelve	and	hundred	twenty
hundred	ninety	eight	nine	and
one	and	hundred	thirteen	five
hundred	and	seven	forty	ten
eleven	and	two	and	fifteen
and	sixteen	fifty	sixty	eighty
three	fourteen	hundred	four	nineteen
hundred	six	seventeen	thirty	and
hundred	five	and	eighteen	seven
START	nine	hundred	seventy	and

Estimate and then count each group of objects. Write both your estimate and the count in your exercise book.

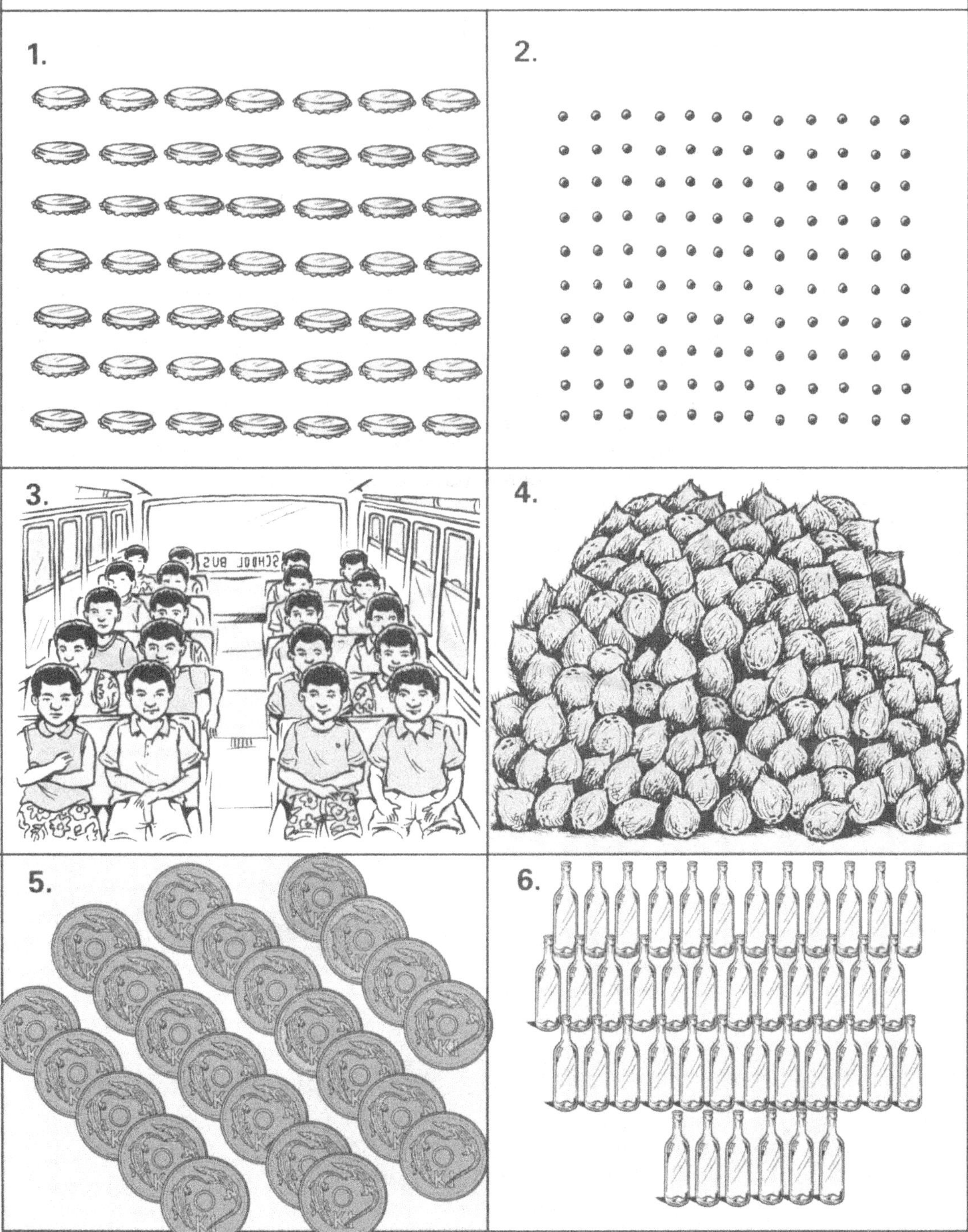

Write number sentences to show how you can use the coins below to make the following amounts. Here is an example.

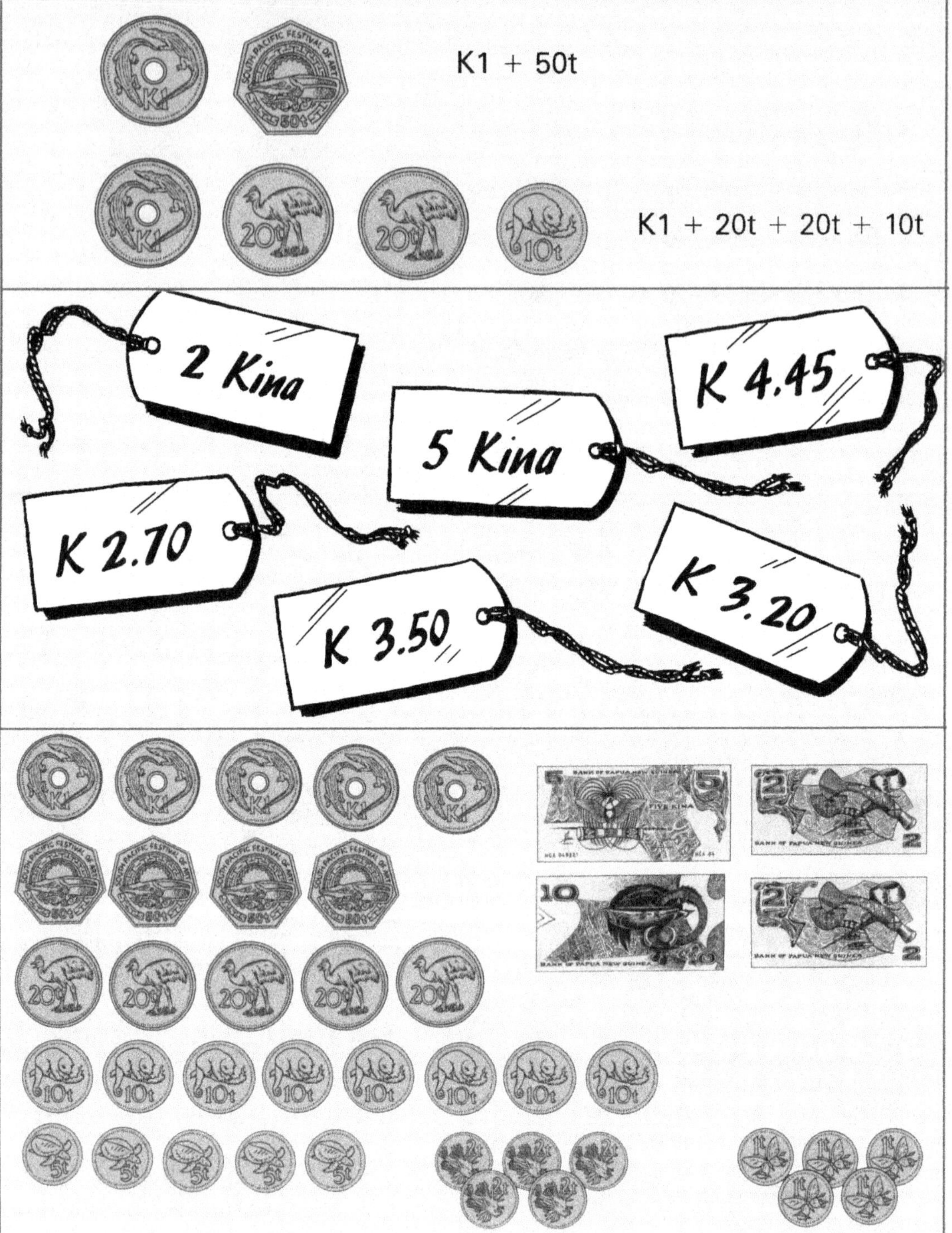

Work out how much change is needed if you buy the items with the notes shown. Write your answer in numbers and words.

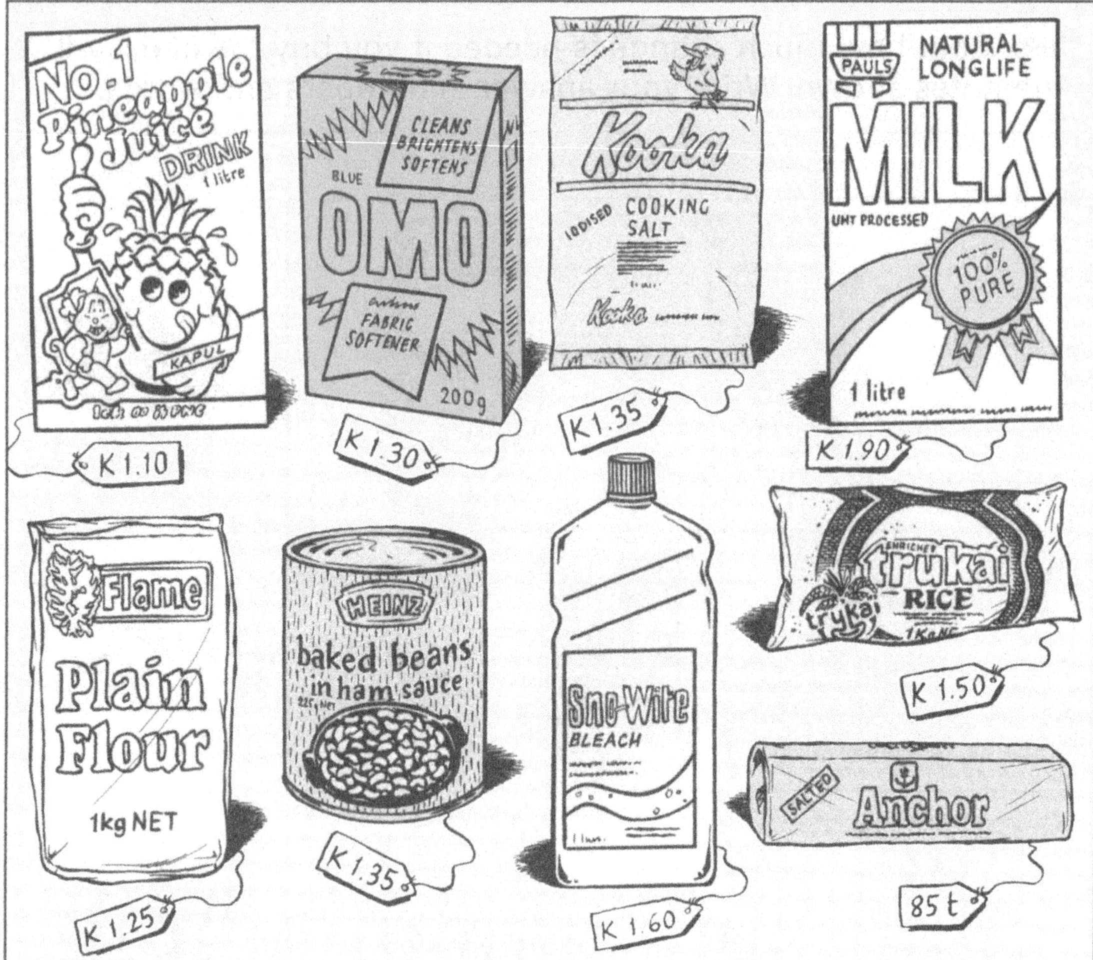

In your exercise book, write the answers to these questions.

1. Which item is cheapest?
2. Which item is most expensive?
3. Find two items under K5.00.
4. Find three items under K7.00.
5. Find two items whose difference is 30t.
6. Which two items cost the same?
7. Make up some number sentences about the prices on this page.

Write your answers to these questions in your exercise book.

1. Which is the cheapest item?
2. How much more than the cheapest item is each of the other items?
3. How much would it cost to buy all items?

Write down your answers in your exercise book.

1. How much would it cost for:
   - a shirt and jogger shoes?
   - a saucepan set and a blanket?
   - a bed sheet, a blanket, and two towels?
   - a wall clock, a set of plates and mugs, and a saucepan set?
2. How much would all of these items cost together?
3. I want to spend K50. What could I buy?
4. I want to spend K100. What could I buy?

A Make clock hands and use them to show different times.

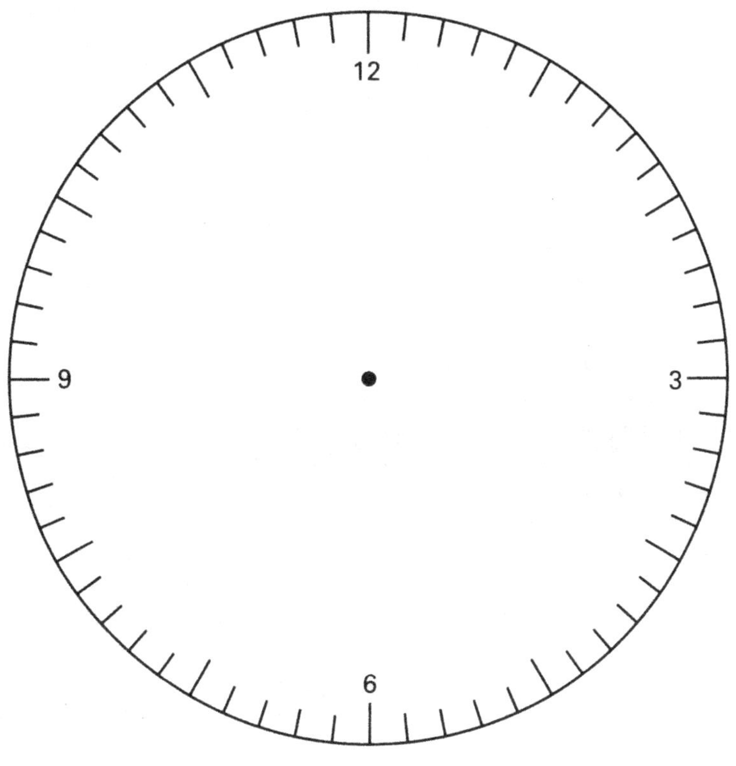

B Write down the time shown on each clock.

C Draw clocks to show these times.

A In your exercise book, write each of these times two ways.

B At what times do each of these events happen?

In your exercise book, draw each of these times onto a clock-face.

A. 8:10

B. 2:20

C. 4:10

D. 11:15

E. 9:55

F. 7:30

G. 8:50

H. 6:40

I. 1:25

J. 3:45

Find the matching times. Copy the digital displays into your exercise book and draw the matching clock-face next to each one.

Use one of these words to say how much time each activity takes.

days	seconds	weeks
minutes	hours	months

1.

2.

3.

4.

5.

6.

7.

8.

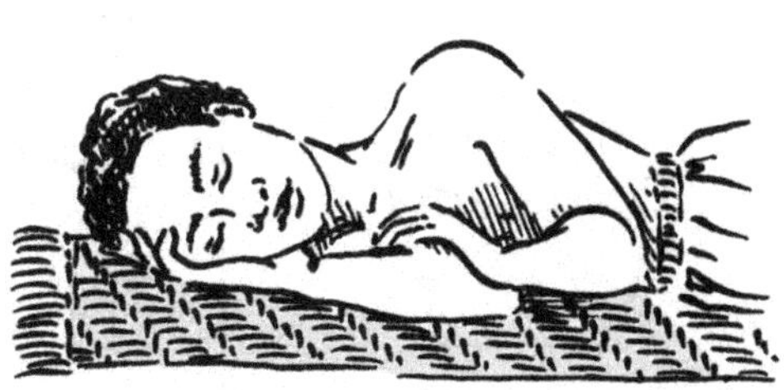

A In your exercise book, draw the diagrams which show $\frac{1}{3}$.

B In your exercise book, write the fractions in symbols and words.

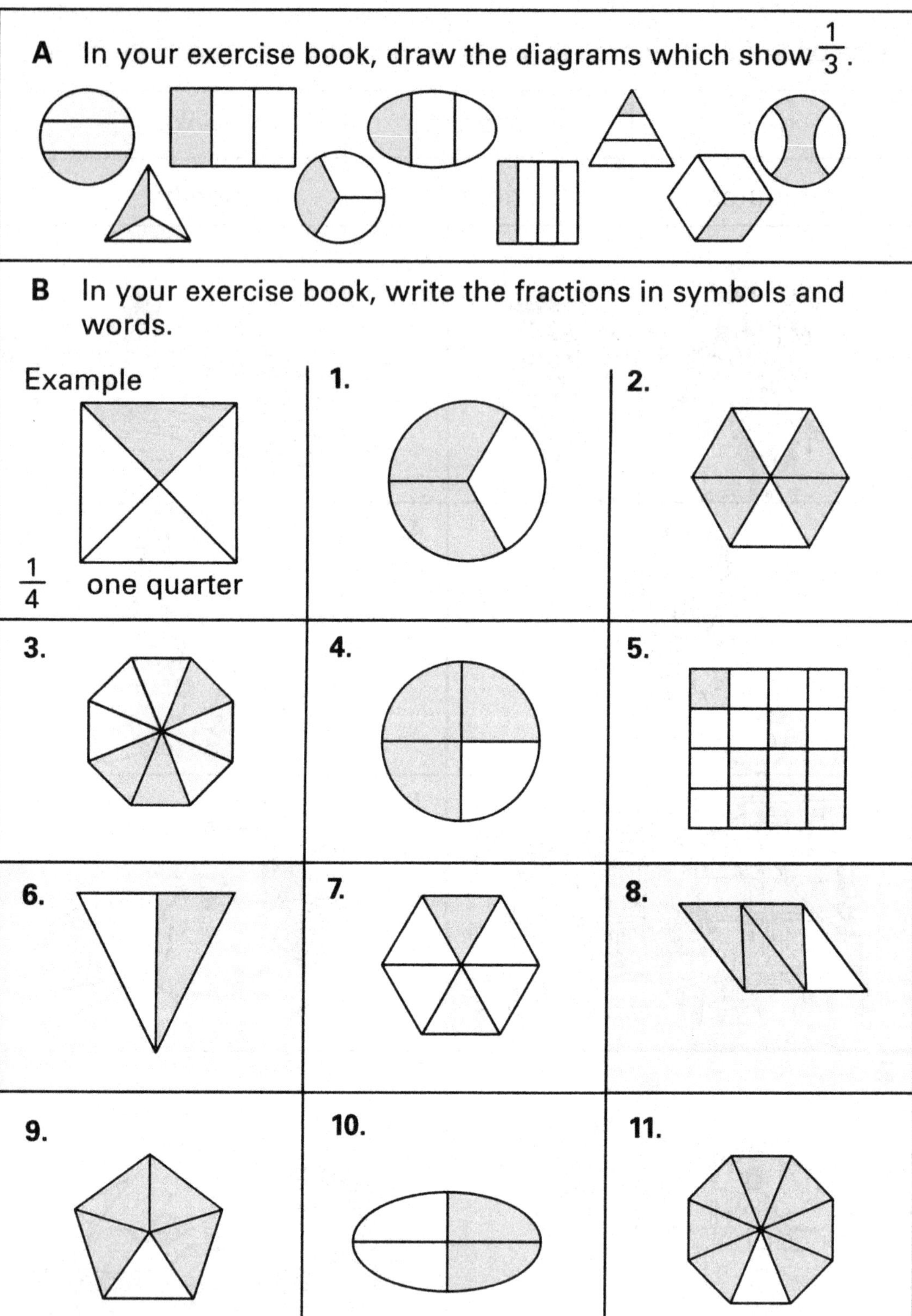

Choose and write the fraction that best shows the shaded part of each drawing.

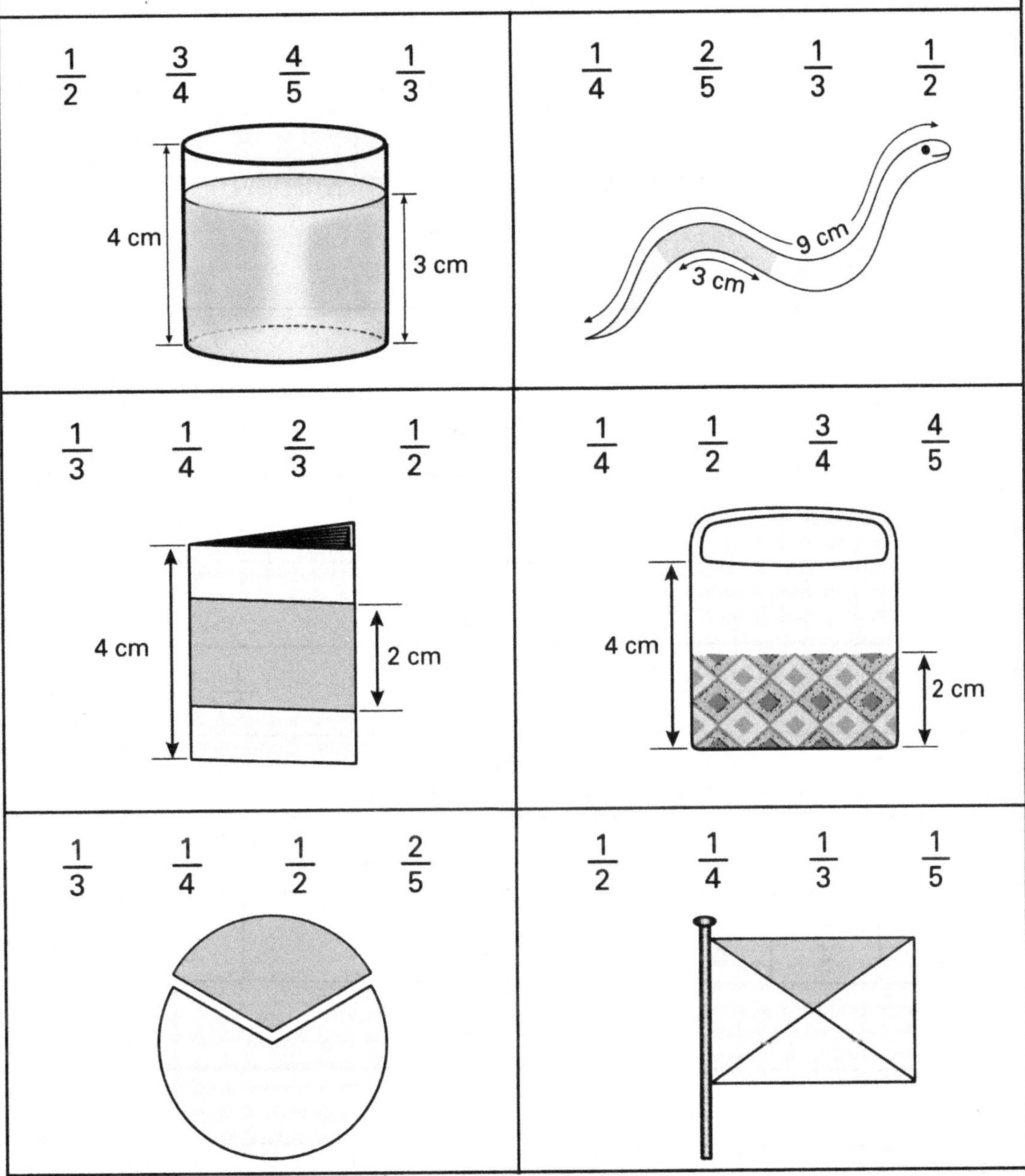

Write each of the fractions in words and draw a different drawing.

What fraction do you need to finish each strip? Write the answer in your exercise book as a number sentence.

Example

$\frac{1}{4}$	$\frac{1}{4}$	

$$\frac{1}{4} + \frac{1}{4} + \frac{1}{2} = 1$$

A.

$\frac{1}{4}$	$\frac{1}{2}$	

B.

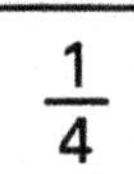

$\frac{1}{4}$	

C.

	$\frac{1}{4}$

D.

$\frac{1}{2}$	

E.

	$\frac{1}{2}$	$\frac{1}{4}$

F.

$\frac{1}{4}$	$\frac{1}{4}$	$\frac{1}{4}$	

G.

	$\frac{1}{4}$	

H.

$\frac{1}{2}$	$\frac{1}{4}$	

I.

	$\frac{1}{2}$	

In your exercise book, write the fraction you need to finish each strip. Write a number sentence. Each strip is worth 1.

a. 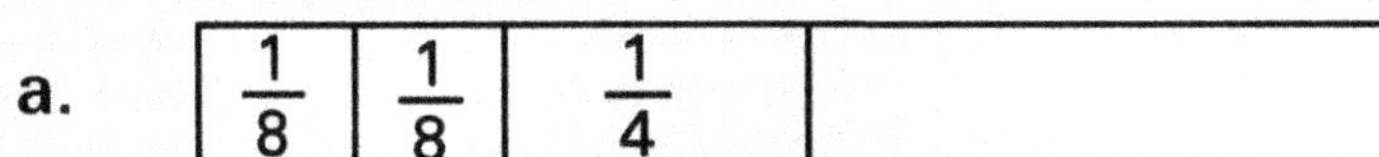

b. $\frac{1}{4}$ $\frac{1}{4}$ $\frac{1}{8}$ $\frac{1}{8}$

c. $\frac{1}{2}$ $\frac{1}{8}$

d. $\frac{1}{8}$ $\frac{1}{2}$ $\frac{1}{4}$

e. $\frac{1}{8}$ $\frac{1}{4}$ $\frac{1}{8}$ $\frac{1}{8}$

f. $\frac{1}{4}$ $\frac{1}{2}$

Will each collection make less than 1, more than 1, or be equal to 1?

1. 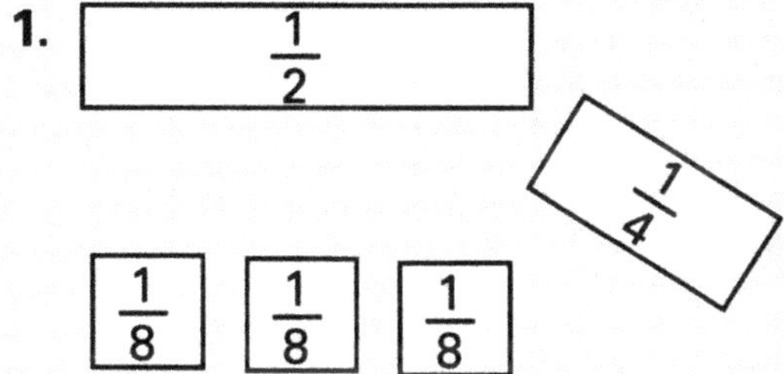

2. 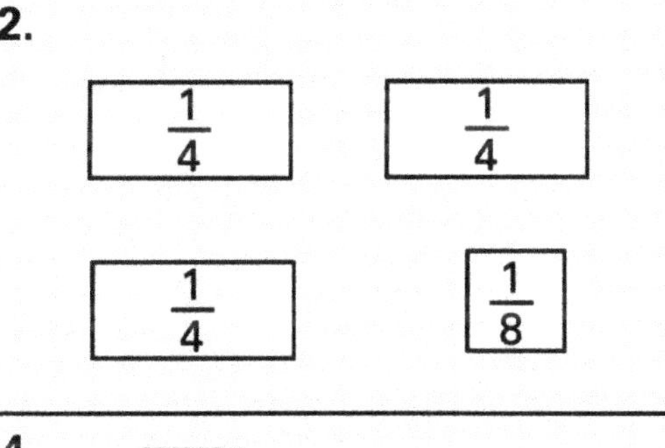

3. 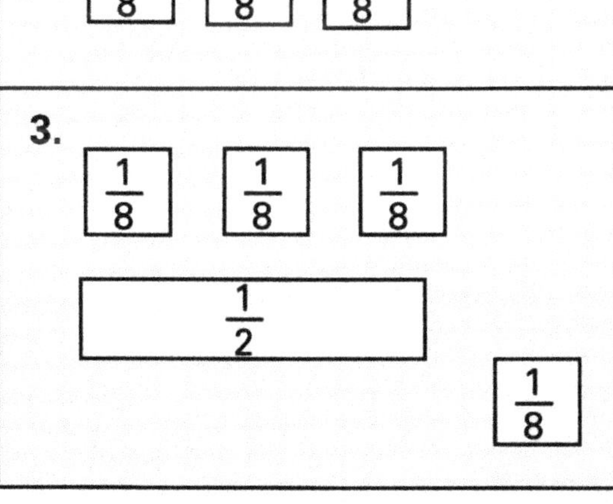

4. 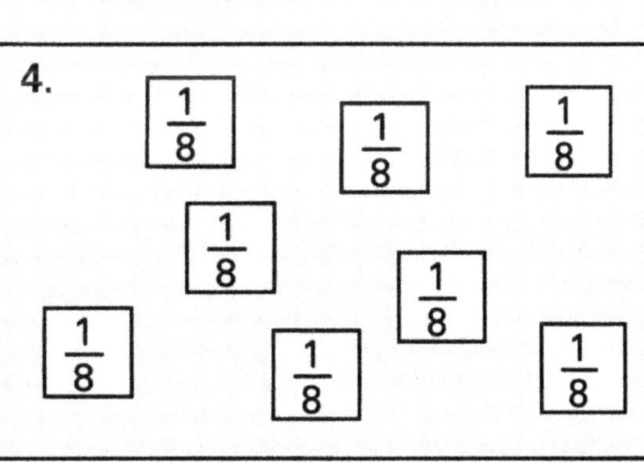

What fraction or mixed number is shown on each number line?

a. 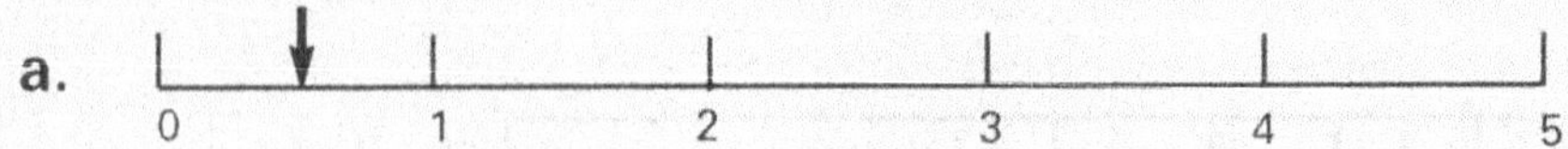

b. 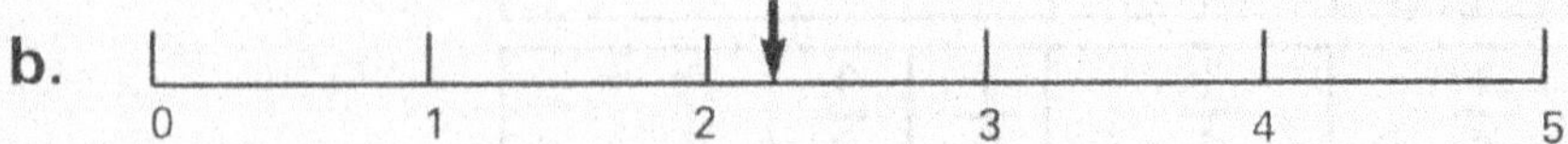

c. 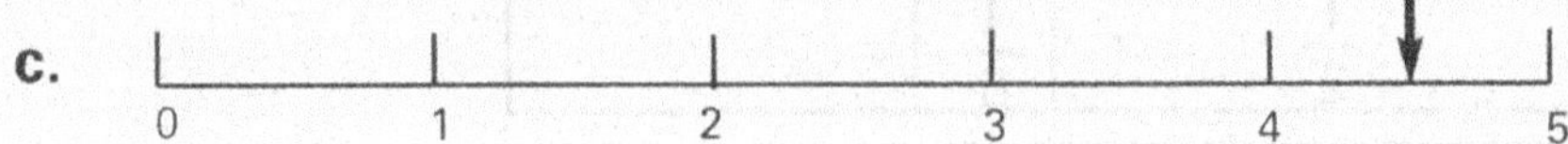

d. 

e. 

f. 

g. 

h. 

Draw a number line for each of these mixed numbers.

$2\frac{1}{2}$ $1\frac{3}{4}$ $4\frac{1}{4}$ $3\frac{1}{2}$

A How many letters are in this sentence?

B Copy these sums into your exercise book. Look for quick ways to do these sums.

a. $5 + 4 + 5 =$

b. $3 + 7 + 7 =$

c. $8 + 5 + 3 =$

d. $2 + 9 + 8 =$

e. $6 + 7 + 4 =$

f. $9 + 6 + 1 =$

g. $\begin{array}{r} 7 \\ 6 \\ +\ 3 \\ \hline \end{array}$

h. $\begin{array}{r} 4 \\ 9 \\ +\ 6 \\ \hline \end{array}$

i. $\begin{array}{r} 8 \\ 4 \\ +\ 8 \\ \hline \end{array}$

j. $\begin{array}{r} 7 \\ 2 \\ +\ 9 \\ \hline \end{array}$

C Copy this table into your exercise book. Add the numbers to fill in the empty boxes. Now add the numbers in the boxes to find what number goes in the circle.

3	3	4	7	□
8	6	3	4	□
2	8	9	6	□
4	3	3	7	□
□	□	□	□	○

Copy and complete these 'magic' squares.

(12)

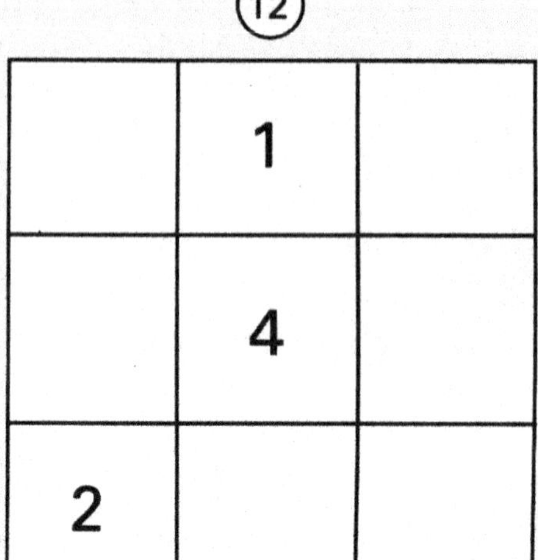

	1	
	4	
2		

(18)

10		7
	6	
	11	

Copy these into your exercise book. Fill in the missing numbers. Check that the totals in the circles are correct.

6	7		5	21
4	5	2	8	
7		4	0	19
1	9	9	6	
	29			(84)

3	0	6	2	
5	5		1	13
	4	3	4	
5	6	0		15
16				(53)

$$\begin{array}{r} 36 \\ +\ {}_{1}17 \\ \hline 53 \end{array}$$

This is the usual way to set out addition sums.

A Copy these sums into your exercise book and work out the answers. Rewrite a. to d. in vertical form.

a. 39 + 5 =

b. 28 + 16 =

c. 35 + 26 =

d. 18 + 38 =

e. $\begin{array}{r} 24 \\ 18 \\ +\ 7 \\ \hline \end{array}$ **f.** $\begin{array}{r} 19 \\ 26 \\ +\ 21 \\ \hline \end{array}$ **g.** $\begin{array}{r} 9 \\ 25 \\ +\ 32 \\ \hline \end{array}$ **h.** $\begin{array}{r} 35 \\ 37 \\ +\ 23 \\ \hline \end{array}$

B Copy these sums into your exercise book and work out the answers. Rewrite a. to f. in vertical form.

a. __ + __ = 35

b. __ + __ = 38

c. __ + __ = 27

d. __ + __ = 41

e. __ + __ + __ = 29

f. __ + __ + __ = 45

g. $\begin{array}{r} \square\square \\ +\ \square\square \\ \hline 3\ 3 \end{array}$ **h.** $\begin{array}{r} \square \\ +\ \square\square \\ \hline 2\ 6 \end{array}$ **i.** $\begin{array}{r} \square \\ \boxed{7} \\ +\ \square\square \\ \hline 4\ 2 \end{array}$ **j.** $\begin{array}{r} \square\square \\ \boxed{2}\boxed{6} \\ +\ \square \\ \hline 5\ 1 \end{array}$

$$\begin{array}{r} 457 \\ +\ {}_{1}1{}_{1}68 \\ \hline 625 \\ \hline \end{array}$$

This is the usual way to set out addition sums.

Copy these sums into your exercise book and work out the answers.

a. $\begin{array}{r} 235 \\ +\ 112 \\ \hline \end{array}$

b. $\begin{array}{r} 126 \\ +\ 64 \\ \hline \end{array}$

c. $\begin{array}{r} 247 \\ +\ 145 \\ \hline \end{array}$

d. $\begin{array}{r} 67 \\ +\ 215 \\ \hline \end{array}$

e. $\begin{array}{r} 35 \\ +\ 253 \\ \hline \end{array}$

f. $\begin{array}{r} 526 \\ +\ 297 \\ \hline \end{array}$

g. $\begin{array}{r} 157 \\ 226 \\ +\ 32 \\ \hline \end{array}$

h. $\begin{array}{r} 96 \\ 88 \\ +\ 257 \\ \hline \end{array}$

i. $\begin{array}{r} 68 \\ 307 \\ +\ 256 \\ \hline \end{array}$

j. $54 + 29 + 276 =$

k. $215 + 176 + 133 =$

l. $98 + 250 + 165 =$

m. $129 + 76 + 253 =$

n. $309 + 67 + 154 =$

Work out the answers to these questions in your exercise book.

1. How far is it by road from Town A to Town C?

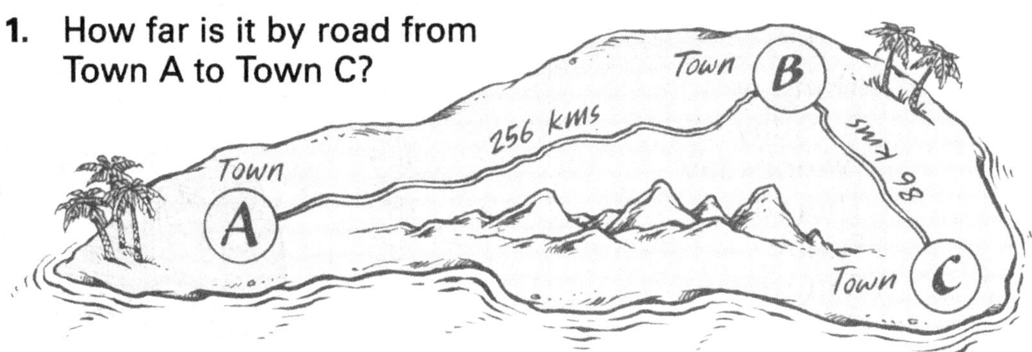

2. A person collects 32 coconuts one day, 56 another day and 71 coconuts the next day. How many coconuts is this altogether?

3. On the first bus load a driver carries 17 passengers, on the second load he carries 24 passengers and on the third load he carries 18 passengers. How many passengers is this altogether?

4. The catering staff at Air Niugini have to plan how many lunches they need to prepare for the plane journey. The first plane has 53 passengers, the second plane has 27 and the third plane has 136. How many lunches do they need to prepare?

5. 127 + 49 + 156 =

Write a story about this sum and then work out the answer.

Write the answers in your exercise book.

A

DAY	Number of kilometres
Monday	134
Tuesday	98
Wednesday	257
Thursday	182
Friday	195
Saturday	87
Sunday	93

1. How many kilometres did the bus travel on Saturday and Sunday?
2. What was the total amount of kilometres travelled on Monday, Tuesday and Wednesday?
3. Did the bus travel further on Tuesday and Wednesday than on Thursday and Friday?

B

1. Two schools together have 451 students. How many students might there be in each of the schools?
2. A village hall was used for a three-day concert. 235 people came over the three days. How many people might have come each day?
3. Over two days the tourist office received 165 phone calls. How many calls might it have received each day?

Which of these things **will** happen?
Which of these things **won't** happen?
Which of these things **might** happen?

Draw a picture of an event which might happen.

Which is more likely?

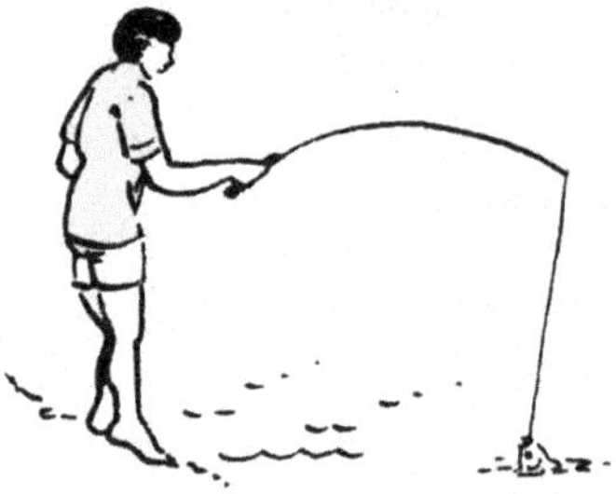

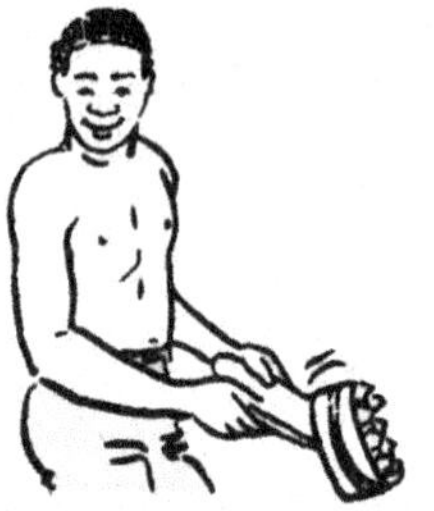

A What could happen next?

B These children could be boys or girls. List all the possibilities.

A Which coins show heads? Which coins show tails?

B What does this table show us?

		Total
Heads	~~1111~~ ~~1111~~ ~~1111~~ 111 ~~1111~~	
Tails	~~1111~~ ~~1111~~ ~~1111~~ ~~1111~~ ~~1111~~ 11	

C Use this table to show what happens when you toss two coins together.

	Tally	Total
2 Heads		
2 Tails		
Head/Tail		
		20

A Copy the chart into your exercise book. Roll two dice and record the total on the chart each time.

Number of dice	Tally	Total
1		
2		
3		
4		
5		
6		
7		
8		
9		
10		
11		
12		
		100

B Who won this dice game?

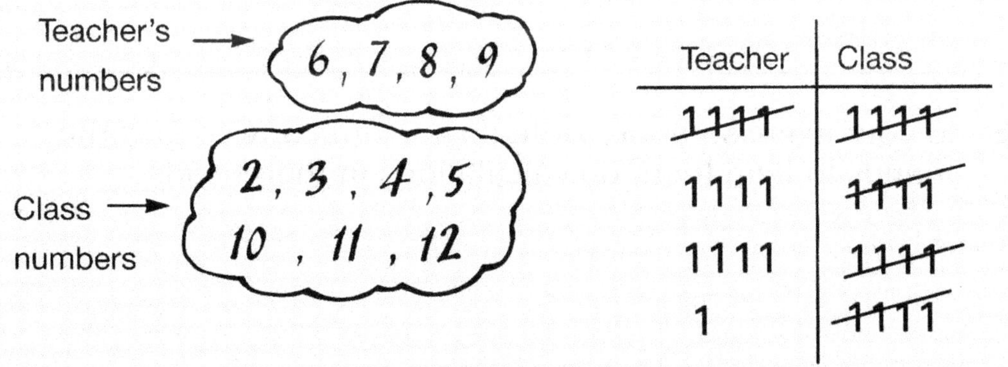

Do all of these in your head.

A

1. $25 + 25$
2. $25 + 26$
3. $24 + 27$
4. $25 + 27$
5. $27 + 27$

B

1. $20 + 20 + 20 + 20$
2. $19 + 19 + 19 + 19$
3. $20 + 21 + 20 + 21$
4. $22 + 22 + 22 + 22$
5. $19 + 22 + 19 + 22$

C

1. $75 - 16$
2. $75 - 17$
3. $75 - 27$
4. $75 - 28$
5. $75 - 32$

D

1. 13×3
2. 23×3
3. 14×3
4. 24×3
5. 15×3

E In your exercise book, write down some advice to some friends to help them do calculations in their heads.

Copy these squares into your exercise book, and fill in the missing numbers to make them magic squares.

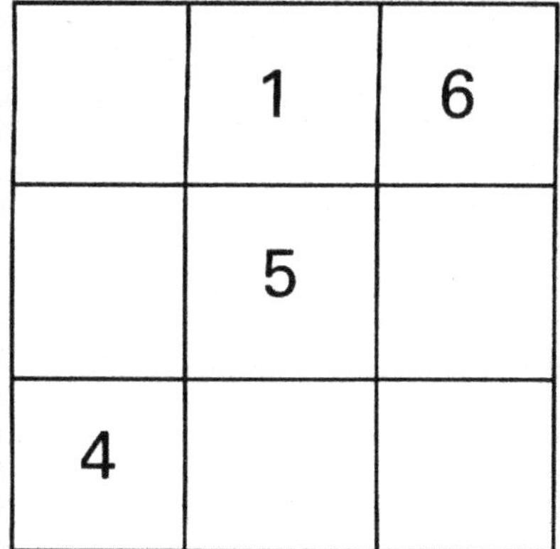

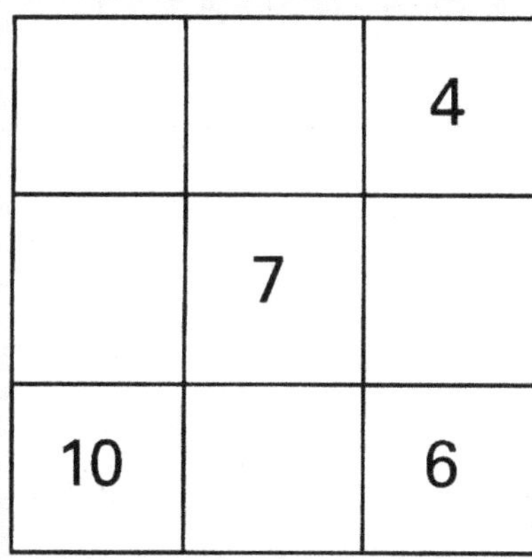

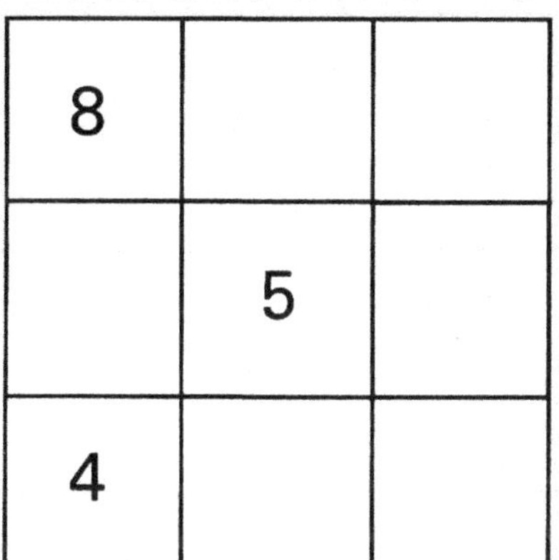

Is this a magic square?

Make up another magic square this size.

16	3	2	13
5	10	11	8
9	6	7	12
4	15	14	1

In your head, work out the answer for each of these sums. Tell a friend how you did each one.	
1. $7 + 3 + 6 + 4 + 5 + 5 + 2 =$	
2. $2 + 4 + 5 + 8 + 6 + 5 + 9 + 1 =$	
3. $40 + 60 + 20 + 70 + 80 + 30 + 6 =$	
4. $50 + 20 + 40 + 80 + 50 + 30 =$	
5. $25 + 30 + 25 + 70 + 25 =$	
6. $50 + 75 + 25 + 75 + 50 + 25 =$	
7. $81 - 56 =$	**9.** $6 \times 27 =$
8. $9 \times 39 =$	**10.** $75 - 39 =$

In your head, work out the answers and write them in your exercise book.

a. $9 - 5 =$

b. $12 - 7 =$

c. $21 - 6 =$

d. $34 - 7 =$

e. $65 - 9 =$

f. $83 - 8 =$

g. $53 - 7 =$

h. $44 - 9 =$

i. $\begin{array}{r} 33 \\ -\ 6 \\ \hline \end{array}$

j. $\begin{array}{r} 46 \\ -\ 8 \\ \hline \end{array}$

k. $\begin{array}{r} 55 \\ -\ 9 \\ \hline \end{array}$

l. $\begin{array}{r} 93 \\ -\ 7 \\ \hline \end{array}$

m. $\begin{array}{r} 63 \\ -\ 5 \\ \hline \end{array}$

n. $\begin{array}{r} 54 \\ -\ 8 \\ \hline \end{array}$

o. $\begin{array}{r} 53 \\ -\ 8 \\ \hline \end{array}$

p. $\begin{array}{r} 34 \\ -\ 5 \\ \hline \end{array}$

q. $\begin{array}{r} 72 \\ -\ 8 \\ \hline \end{array}$

$$\begin{array}{r} \overset{2}{\cancel{3}}\overset{15}{\cancel{6}}\overset{1}{2} \\ -\ 175 \\ \hline 187 \end{array}$$

This is the usual way to set out subtractions.

Copy and complete.

a. $\begin{array}{r} 53 \\ -\ 25 \\ \hline \end{array}$ b. $\begin{array}{r} 87 \\ -\ 34 \\ \hline \end{array}$ c. $\begin{array}{r} 75 \\ -\ 18 \\ \hline \end{array}$

d. $\begin{array}{r} 76 \\ -\ 20 \\ \hline \end{array}$ e. $\begin{array}{r} 80 \\ -\ 17 \\ \hline \end{array}$ f. $\begin{array}{r} 82 \\ -\ 66 \\ \hline \end{array}$

g. $\begin{array}{r} 445 \\ -\ 267 \\ \hline \end{array}$ h. $\begin{array}{r} 514 \\ -\ 148 \\ \hline \end{array}$ i. $\begin{array}{r} 237 \\ -\ 59 \\ \hline \end{array}$

j. $\begin{array}{r} 362 \\ -\ 217 \\ \hline \end{array}$ k. $\begin{array}{r} 615 \\ -\ 43 \\ \hline \end{array}$ l. $\begin{array}{r} 842 \\ -\ 129 \\ \hline \end{array}$

m. $\begin{array}{r} 274 \\ -\ 190 \\ \hline \end{array}$ n. $\begin{array}{r} 372 \\ -\ 206 \\ \hline \end{array}$ o. $\begin{array}{r} 927 \\ -\ 360 \\ \hline \end{array}$

p. $\begin{array}{r} 342 \\ -\ 277 \\ \hline \end{array}$ q. $\begin{array}{r} 512 \\ -\ 177 \\ \hline \end{array}$ r. $\begin{array}{r} 418 \\ -\ 109 \\ \hline \end{array}$

Furniture	Usual Price	Sale Price
Washing Machine	K692	K499
Refrigerator	K739	K549
Television	K545	K489
Armchair	K135	K89
Esky	K246	K165
Table and Chairs	K449	K385

a. How much money could I save on each item?

b. How much would I save if I bought an esky and two armchairs?

c. How much would I save if I bought the table and chairs and the refrigerator?

d. How much would I save if I bought one of each of the items on sale?

e. How much more is a television than an esky?

f. How much more is a washing machine than an armchair?

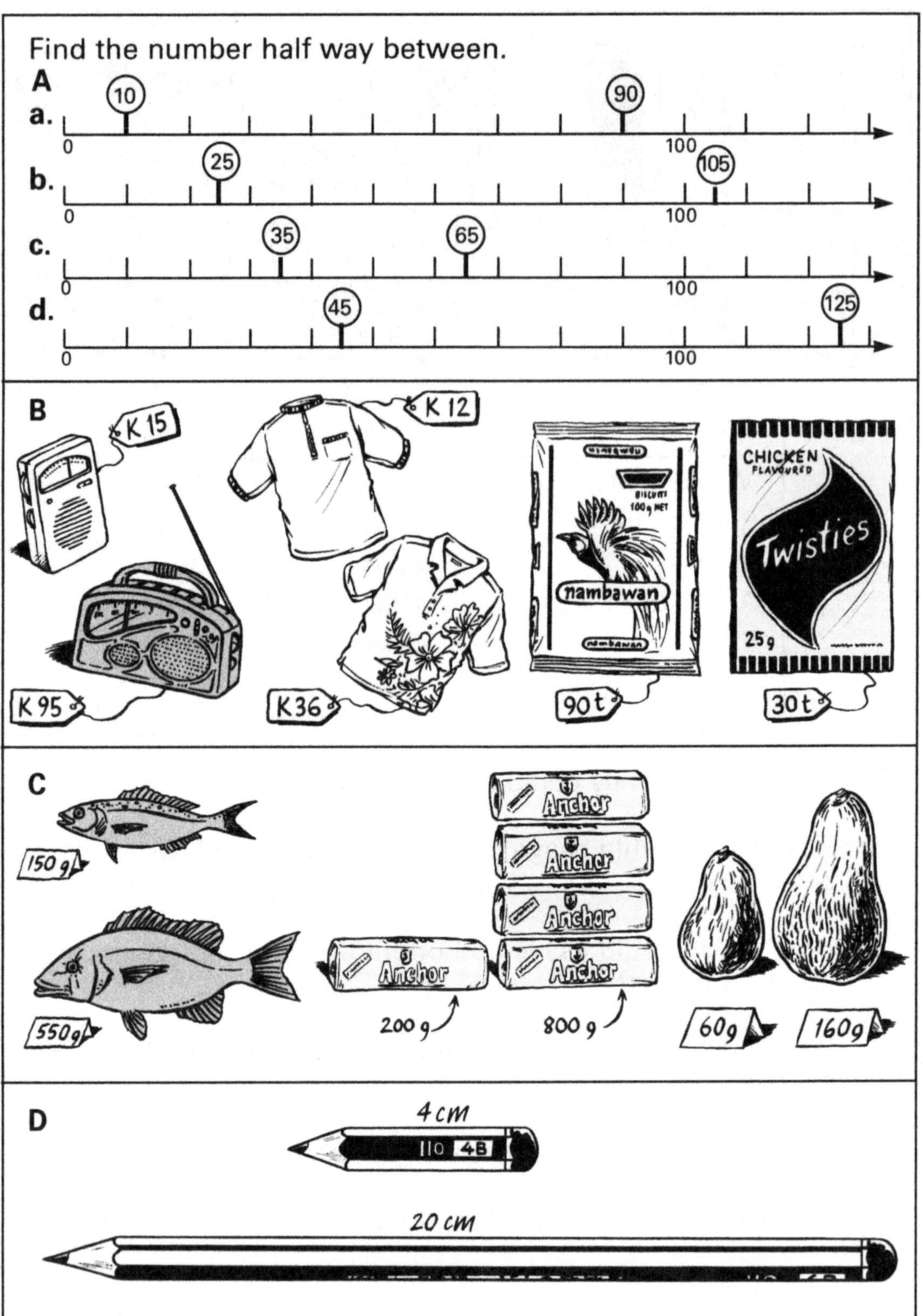
Find the number half way between.
A
a.
10
90
0
100
b.
25
105
0
100
c.
35
65
0
100
d.
45
125
0
100
B
K 15
K 12
K 95
K36
CHICKEN
FLAVOURED
Twisties
25g
nambawan
90t
30t
C
150 g
550g
Anchor
200 g
800 g
60g
160g
D
4 cm
4B
20 cm

Subtracting with zero

$$\begin{array}{r} {}^{5}\ {}^{1} \\ 3\not{6}0 \\ -\ 147 \\ \hline 213 \end{array} \qquad \begin{array}{r} {}^{9} \\ {}^{3}\ {}^{1}\ {}^{1} \\ \not{4}\not{0}5 \\ -\ 238 \\ \hline 167 \end{array}$$

a. $\begin{array}{r} 180 \\ -\ 63 \\ \hline \end{array}$

b. $\begin{array}{r} 270 \\ -\ 155 \\ \hline \end{array}$

c. $\begin{array}{r} 380 \\ -\ 68 \\ \hline \end{array}$

d. $\begin{array}{r} 580 \\ -\ 158 \\ \hline \end{array}$

e. $\begin{array}{r} 390 \\ -\ 149 \\ \hline \end{array}$

f. $\begin{array}{r} 490 \\ -\ 165 \\ \hline \end{array}$

g. $\begin{array}{r} 608 \\ -\ 247 \\ \hline \end{array}$

h. $\begin{array}{r} 805 \\ -\ 654 \\ \hline \end{array}$

i. $\begin{array}{r} 903 \\ -\ 180 \\ \hline \end{array}$

j. $\begin{array}{r} 702 \\ -\ 155 \\ \hline \end{array}$

k. $\begin{array}{r} 305 \\ -\ 147 \\ \hline \end{array}$

l. $\begin{array}{r} 408 \\ -\ 169 \\ \hline \end{array}$

In your exercise book, draw a picture for each story, and solve the problem.

1. On Thursday, 125 children were at school. 64 were boys. How many were girls?

2. On Friday, 125 children came to school and 86 went to play sport with another school. How many children were left at school?

3. The trade store has 371 tins that are either meat or fish. I counted 159 tins of meat. How many tins of fish are there?

4. After the rain I checked my stack of pineapples. Out of 367, 158 of them had gone bad. How many good ones were left?

5. We collected 340 coconuts. There are still 207 left. How many have we used?

Starting from the same position, what turns does this person have to make, to face each side of the room?

Which things are half a turn apart?

Find the 'right' or 'quarter turn' angles.

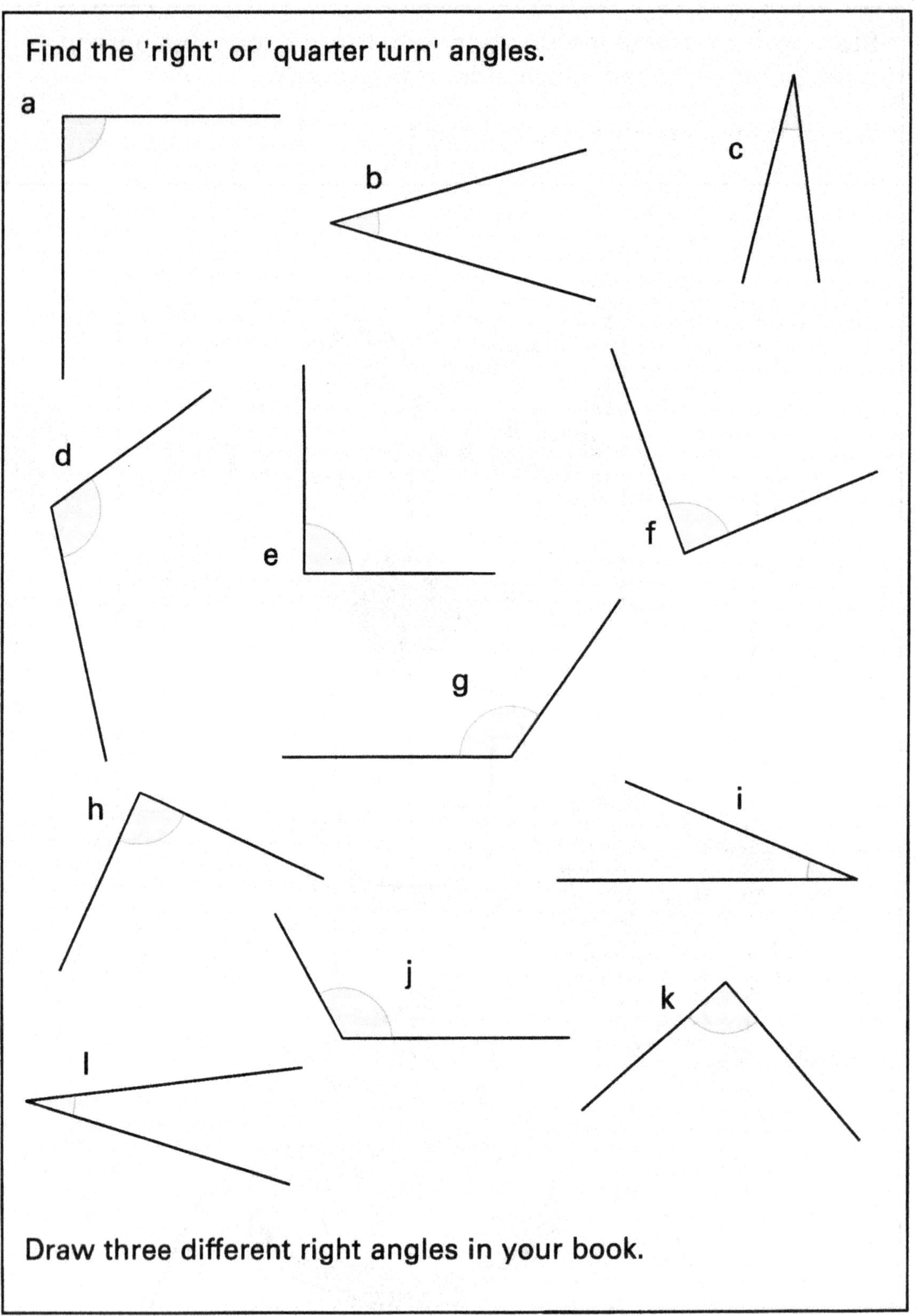

Draw three different right angles in your book.

1. In your exercise book, copy the shapes that are bigger than right angles. Mark these angles.

2. Copy the shapes that have angles less than right angles. Mark these angles.

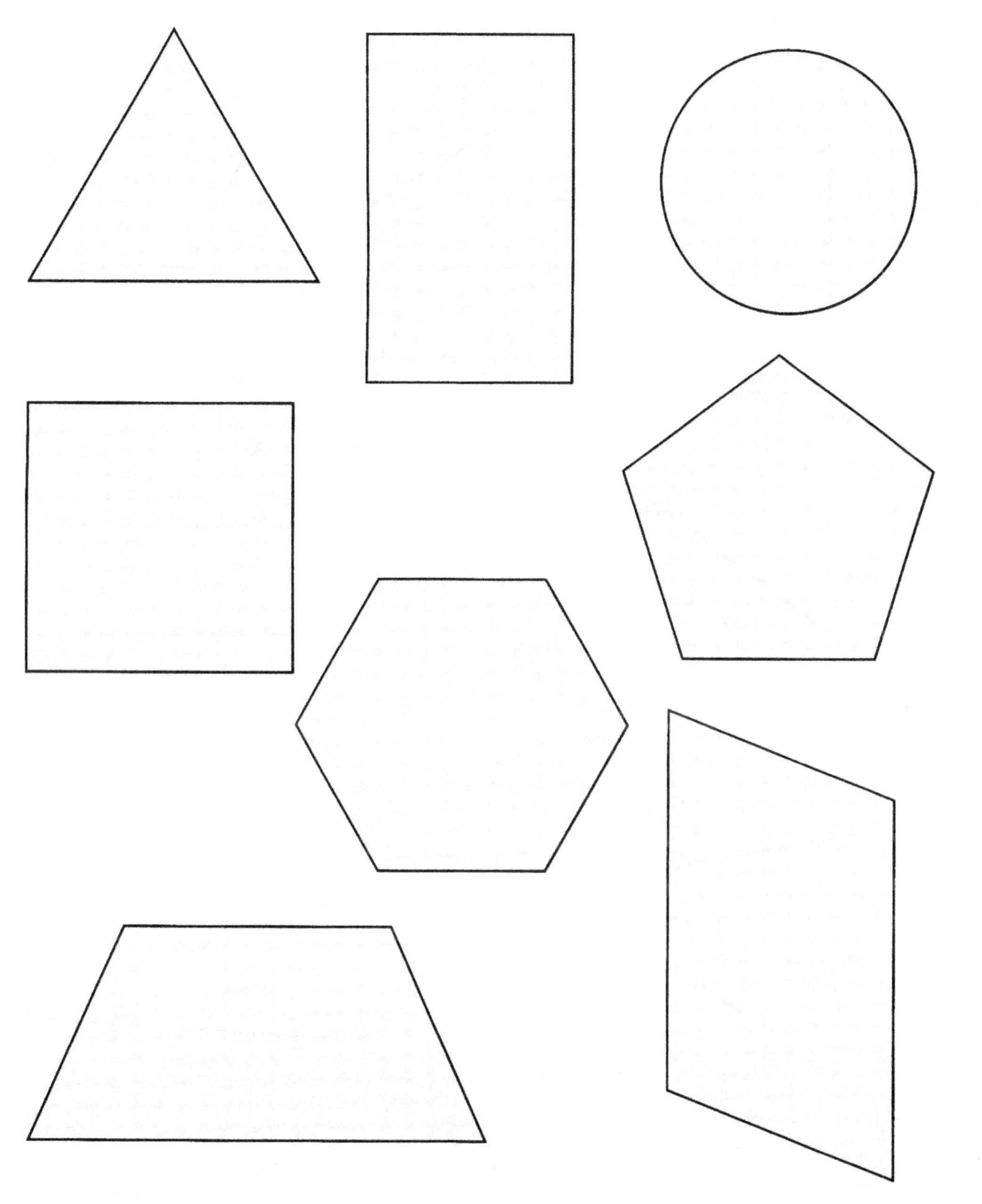

In your exercise book, write the letters of the angles from smallest to largest.

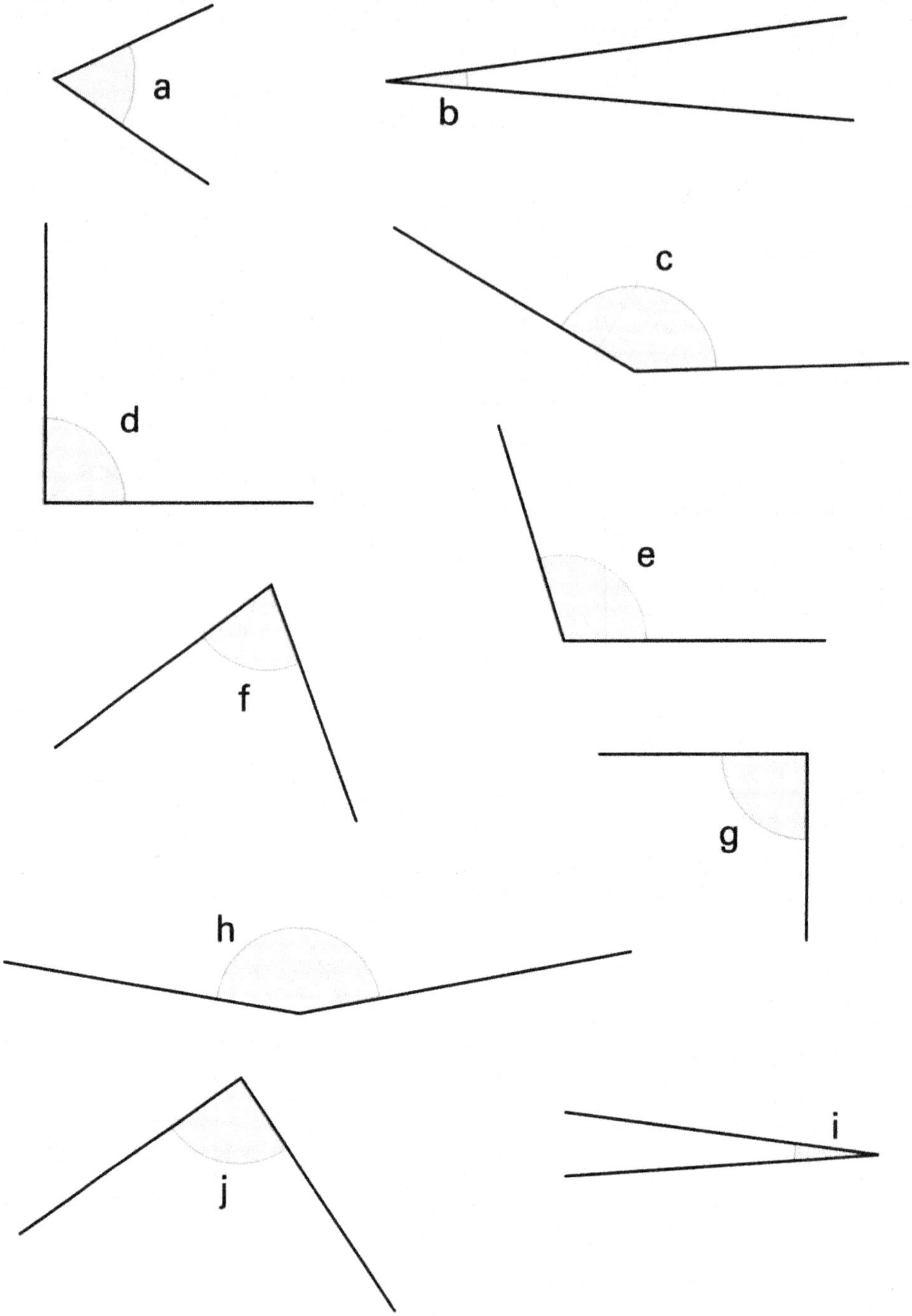

a

b

c

d

e

f

g

h

i

j

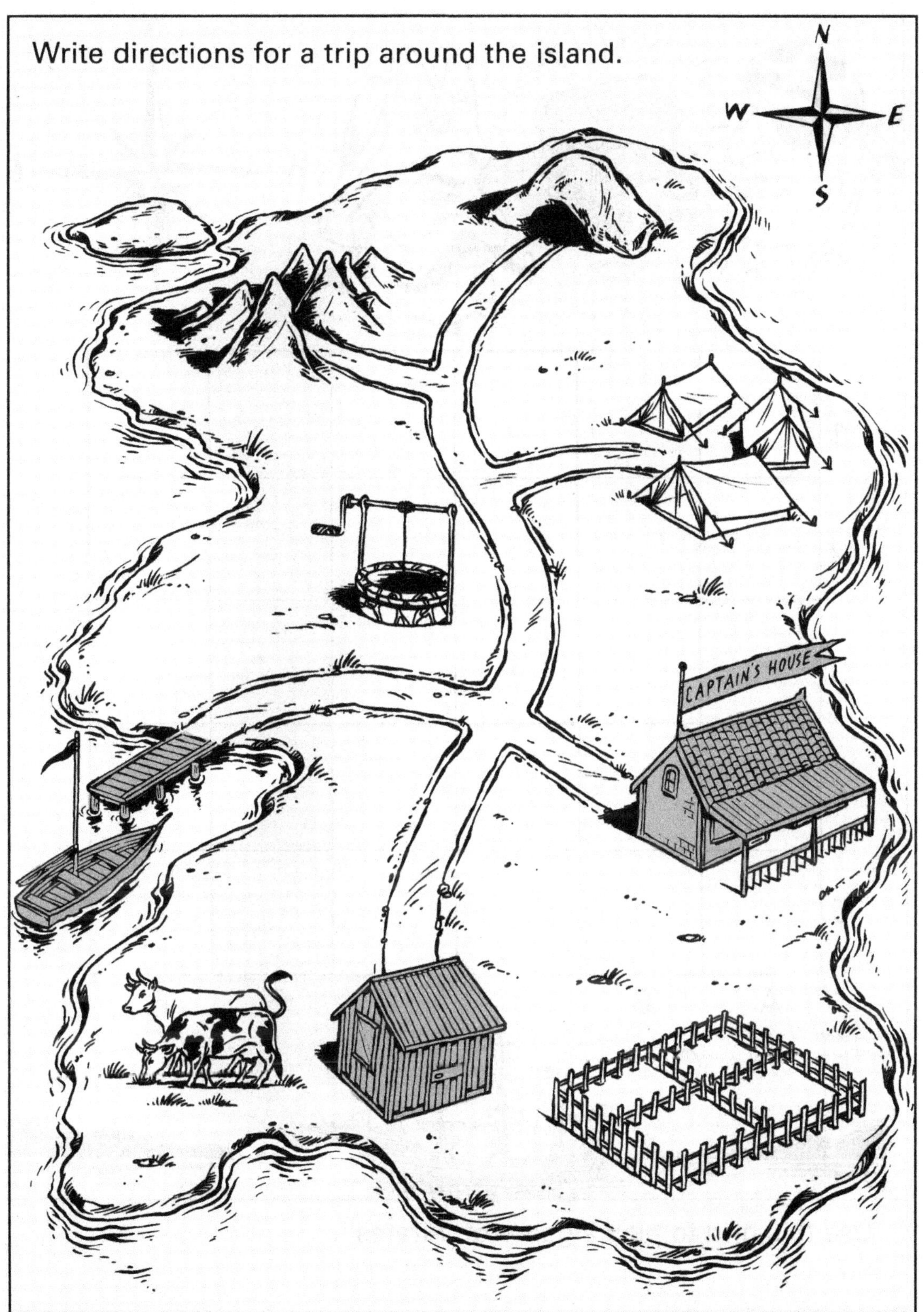
Write directions for a trip around the island.
N
W
E
S
CAPTAIN'S HOUSE

Use this grid to play a game of 'Pirates'.

This is a plan of a school. If you start at the front gate what directions will you walk to reach the fruit trees?

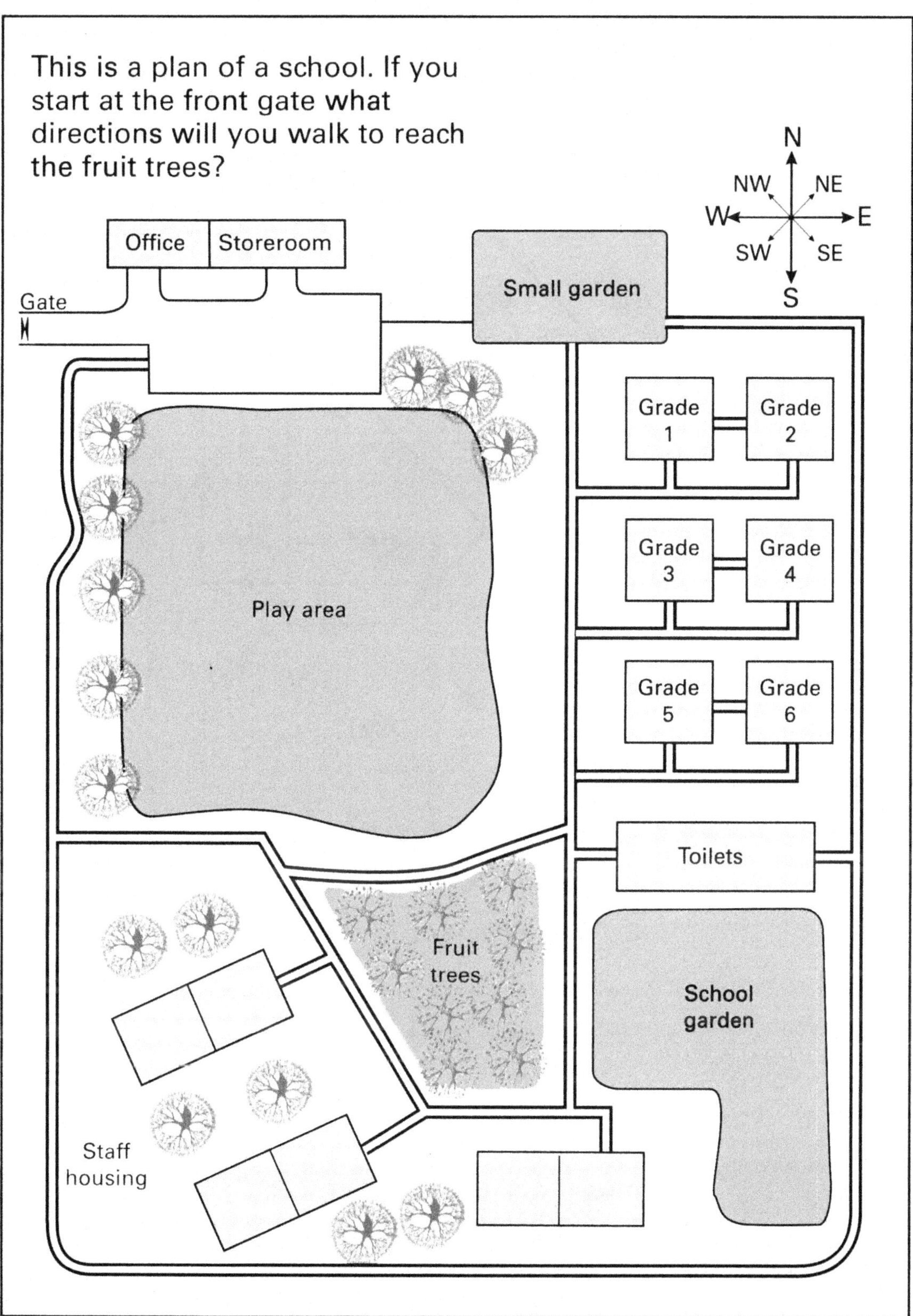

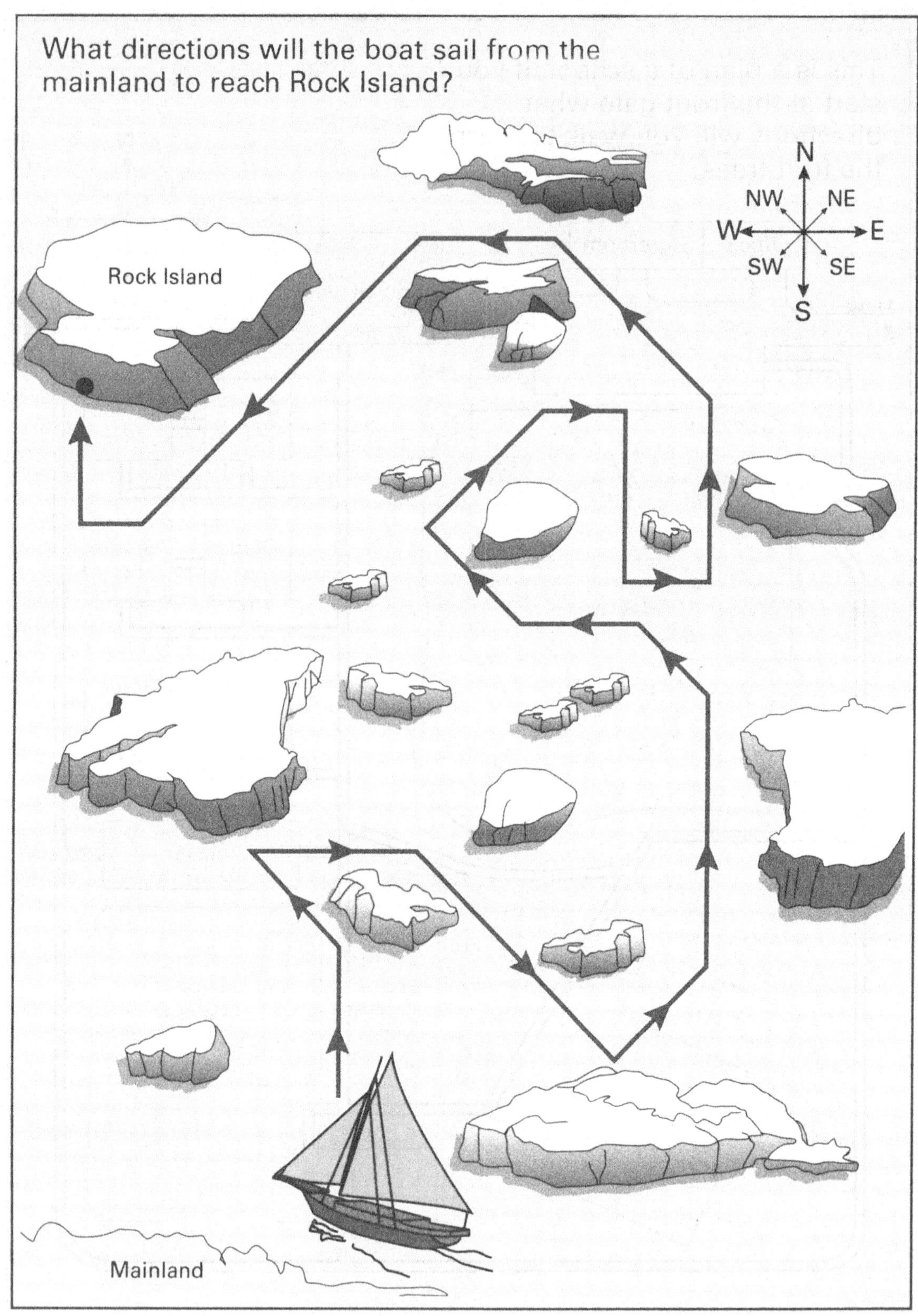
What directions will the boat sail from the mainland to reach Rock Island?
N
NW
NE
W
E
SW
SE
S
Rock Island
Mainland

The Six Times Table

Write a table fact for each picture.

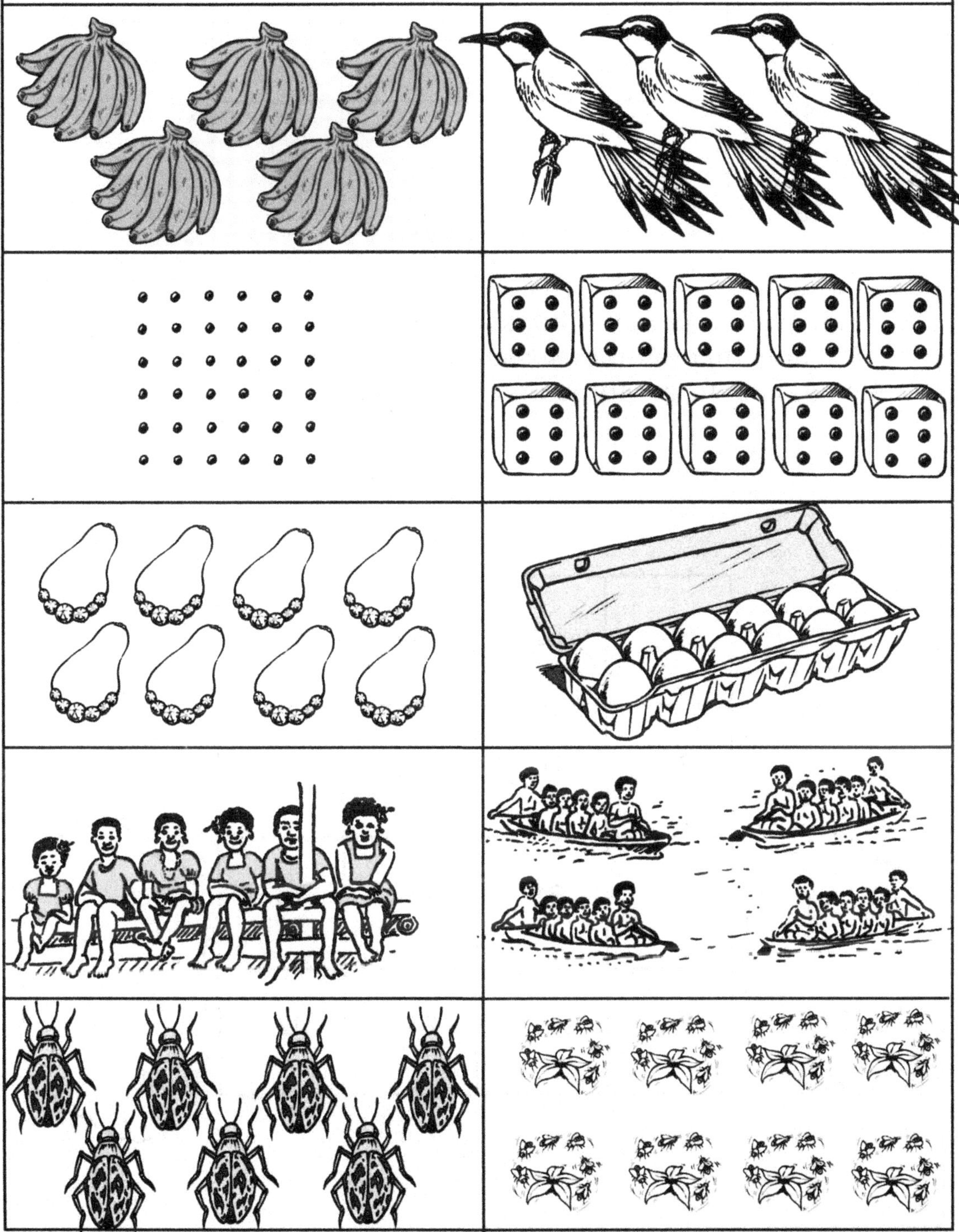

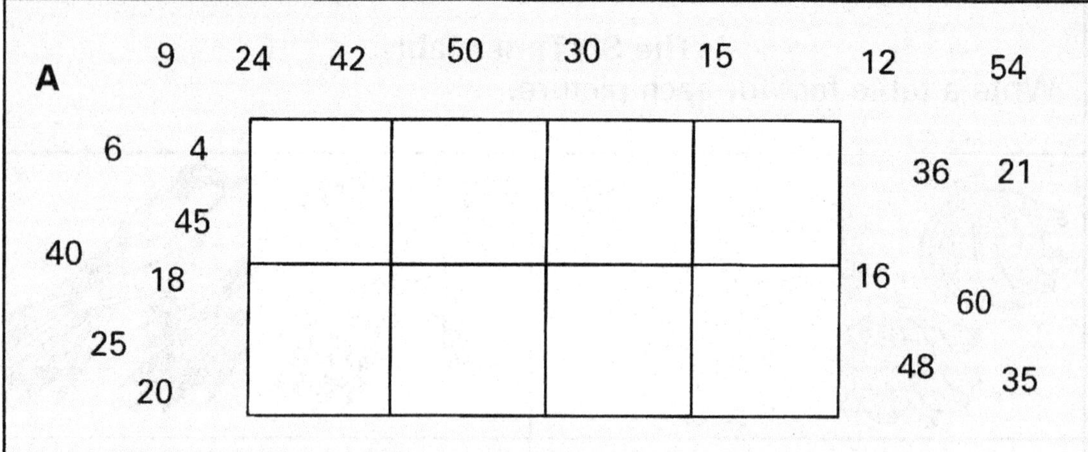

B Play with a partner. Choose a number from each cloud. Multiply them. If the answer is on the grid, place a counter on it. First to get three in a row wins.

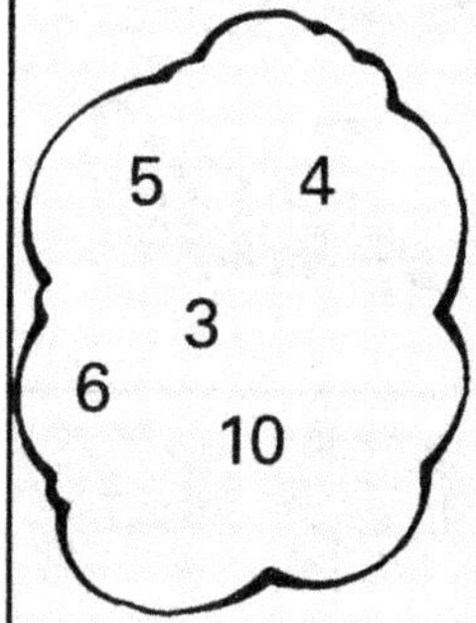

35	50	30	16
32	20	25	12
18	21	28	15
36	40	24	42

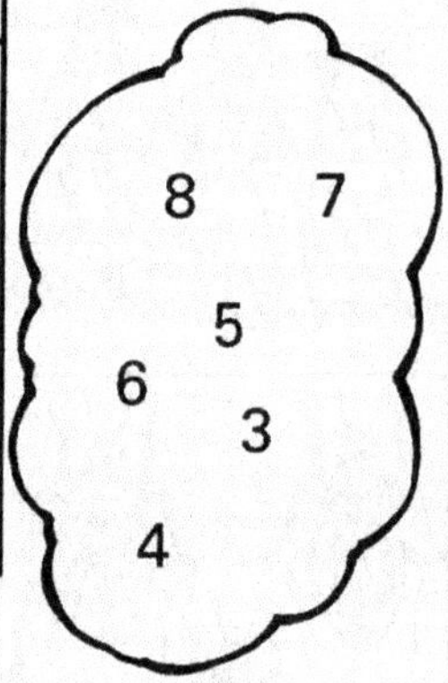

5 4 3 6 10

8 7 5 6 3 4

To play this game you will need 3 friends, and a dice. Each person needs to have a copy of this board and some counters.

9	6	50	30	4	15
18	15	36	24	12	20
12	40	16	2	5	8
60	25	8	50	3	18
10	4	30	36	6	10
24	16	2	20	5	9

Make four of each of the following cards x2, x3, x4, x5, x6, and x10. Mix up the cards and place them face down in the middle of the group. The first player picks up the top card and rolls the dice. He/she should multiply the number on the dice by the number on the card. For example, if the dice is 6 and the card is x10, then the number will be 60. The person should find 60 on their board and cover it with a counter. Once a counter is placed it cannot be moved. The winner is the first person to get four numbers in a row either horizontally, vertically or diagonally.

A

$$\begin{array}{r} 23 \\ \times 3 \\ \hline 69 \end{array} \qquad \begin{array}{r} {}^{1}26 \\ \times 3 \\ \hline 78 \end{array}$$

B Copy these sums into your exercise book and work out the answers.

a. $\begin{array}{r} 43 \\ \times 2 \\ \hline \end{array}$ b. $\begin{array}{r} 24 \\ \times 2 \\ \hline \end{array}$ c. $\begin{array}{r} 31 \\ \times 3 \\ \hline \end{array}$

d. $\begin{array}{r} 16 \\ \times 4 \\ \hline \end{array}$ e. $\begin{array}{r} 21 \\ \times 4 \\ \hline \end{array}$ f. $\begin{array}{r} 27 \\ \times 2 \\ \hline \end{array}$

g. $\begin{array}{r} 46 \\ \times 2 \\ \hline \end{array}$ h. $\begin{array}{r} 27 \\ \times 3 \\ \hline \end{array}$ i. $\begin{array}{r} 23 \\ \times 2 \\ \hline \end{array}$

j. $\begin{array}{r} 18 \\ \times 4 \\ \hline \end{array}$ k. $\begin{array}{r} 25 \\ \times 3 \\ \hline \end{array}$ l. $\begin{array}{r} 13 \\ \times 3 \\ \hline \end{array}$

m. $\begin{array}{r} 15 \\ \times 4 \\ \hline \end{array}$ n. $\begin{array}{r} 23 \\ \times 3 \\ \hline \end{array}$ o. $\begin{array}{r} 26 \\ \times 2 \\ \hline \end{array}$

p. $\begin{array}{r} 26 \\ \times 3 \\ \hline \end{array}$ q. $\begin{array}{r} 14 \\ \times 3 \\ \hline \end{array}$ r. $\begin{array}{r} 13 \\ \times 4 \\ \hline \end{array}$

A Multiplying by Ten

Copy these sums into your exercise book and work out the answers.

a. $6 \times 10 =$ b. $10 \times 4 =$

c. $10 \times 7 =$ d. $5 \times 10 =$

e. $3 \times 10 =$ f. $10 \times 9 =$

g. $10 \times 2 =$ h. $8 \times 10 =$

i. $15 \times 10 =$ j. $10 \times 12 =$

k. $10 \times 10 =$ l. $23 \times 10 =$

m. $38 \times 10 =$ n. $10 \times 18 =$

o. $10 \times 46 =$ p. $13 \times 10 =$

q. $52 \times 10 =$ r. $10 \times 27 =$

s. $31 \times 10 =$ t. $40 \times 10 =$

B What might be the missing numbers?

$\_\_ \times 10 = \_\_\ 0$

Find some examples of mixed numbers in the pictures, e.g. there are $3\frac{1}{2}$ jars of lollies.

A Fraction Wall

1

$\frac{1}{2}$ $\frac{1}{2}$

$\frac{1}{3}$ $\frac{1}{3}$ $\frac{1}{3}$

$\frac{1}{4}$ $\frac{1}{4}$ $\frac{1}{4}$ $\frac{1}{4}$

$\frac{1}{5}$ $\frac{1}{5}$ $\frac{1}{5}$ $\frac{1}{5}$ $\frac{1}{5}$

$\frac{1}{6}$ $\frac{1}{6}$ $\frac{1}{6}$ $\frac{1}{6}$ $\frac{1}{6}$ $\frac{1}{6}$

$\frac{1}{7}$ $\frac{1}{7}$ $\frac{1}{7}$ $\frac{1}{7}$ $\frac{1}{7}$ $\frac{1}{7}$ $\frac{1}{7}$

$\frac{1}{8}$ $\frac{1}{8}$ $\frac{1}{8}$ $\frac{1}{8}$ $\frac{1}{8}$ $\frac{1}{8}$ $\frac{1}{8}$ $\frac{1}{8}$

Which fraction is bigger?

a. $\frac{1}{5}$ or $\frac{1}{8}$

b. $\frac{1}{4}$ or $\frac{1}{7}$

c. $\frac{1}{2}$ or $\frac{1}{6}$

d. $\frac{1}{3}$ or $\frac{1}{4}$

e. $\frac{2}{3}$ or $\frac{2}{4}$

f. $\frac{1}{2}$ or $\frac{3}{5}$

g. $\frac{1}{4}$ or $\frac{3}{6}$

h. $\frac{5}{8}$ or $\frac{2}{6}$

i. $\frac{2}{3}$ or $\frac{6}{7}$

j. $\frac{1}{2}$ or $\frac{4}{7}$

1. Use the fraction wall on page 55 to find a fraction the same size as:

a. $\frac{2}{4}$

b. $\frac{4}{6}$

c. $\frac{1}{3}$

d. $\frac{2}{8}$

e. $\frac{1}{2}$

f. $\frac{3}{4}$

g. $\frac{2}{3}$

h. $\frac{4}{8}$

i. $\frac{3}{6}$

j. $\frac{6}{8}$

2. Find 4 fractions that are bigger than $\frac{1}{3}$.

3. Find 4 fractions that are smaller than $\frac{5}{6}$.

4. What are some fractions bigger than $\frac{1}{3}$ but smaller than $\frac{5}{6}$?

5. What are some fractions bigger than $\frac{1}{4}$ but smaller than $\frac{4}{8}$?

What fraction of the coconuts have fallen?

What fraction of this family are children?

What fraction of the tin fish are left?

What fraction of the children are girls?

What fraction of the children are swimming?

What fraction of the fish are spotted?

What fraction of the baskets are empty?

What fraction of the men are in the boat?

Multiplying by Twenty

$6 \times 20 = 120t$

$6 \times 2 \times 10 = 120t$

Copy these sums into your exercise book and work out the answers.

a. $4 \times 20 =$ **b.** $9 \times 20 =$

c. $20 \times 12 =$ **d.** $20 \times 15 =$

e. $20 \times 16 =$ **f.** $20 \times 18 =$

g. $14 \times 20 =$ **h.** $22 \times 20 =$

i. $20 \times 5 =$ **j.** $20 \times 10 =$

k. $20 \times 25 =$ **l.** $20 \times 20 =$

m. $30 \times 20 =$ **n.** $20 \times 75 =$

o. $20 \times 8 =$ **p.** $6 \times 20 =$

q. $35 \times 20 =$ **r.** $20 \times 45 =$

A

$$\begin{array}{r} 123 \\ \times 3 \\ \hline 369 \end{array} \qquad \begin{array}{r} {}^{1}1{}^{1}45 \\ \times 3 \\ \hline 435 \end{array}$$

B Copy these sums into your exercise book and work out the answers.

a. $\begin{array}{r} 234 \\ \times 2 \\ \hline \end{array}$ **b.** $\begin{array}{r} 115 \\ \times 3 \\ \hline \end{array}$

c. $\begin{array}{r} 245 \\ \times 2 \\ \hline \end{array}$ **d.** $\begin{array}{r} 173 \\ \times 5 \\ \hline \end{array}$

e. $\begin{array}{r} 308 \\ \times 3 \\ \hline \end{array}$ **f.** $\begin{array}{r} 144 \\ \times 3 \\ \hline \end{array}$

g. $234 \times 4 =$ **h.** $176 \times 4 =$

i. $211 \times 5 =$ **j.** $120 \times 5 =$

k. $209 \times 5 =$ **l.** $200 \times 5 =$

1. If I have to build 20 boxes and each box takes me 15 minutes, how many minutes will it take?

2. Twelve trucks arrive, each carrying 20 drums of petrol. How many drums is this?

3. I cut down 5 hands of bananas, each containing 145 bananas. How many bananas is this?

4. If I have 47 twenty toea coins, how much money do I have?

5. How many fingers are there in your classroom right now?

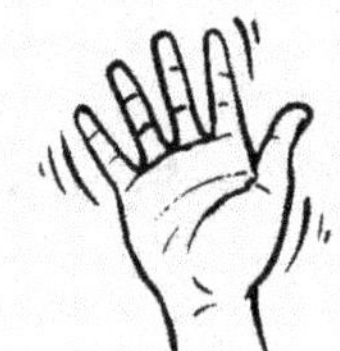

6. How many minutes are there in four hours?

7. How many hours are there in one week?

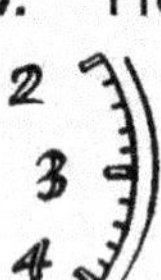

8. How many legs are there on all the tables in your classroom?

A

12 lots of 13

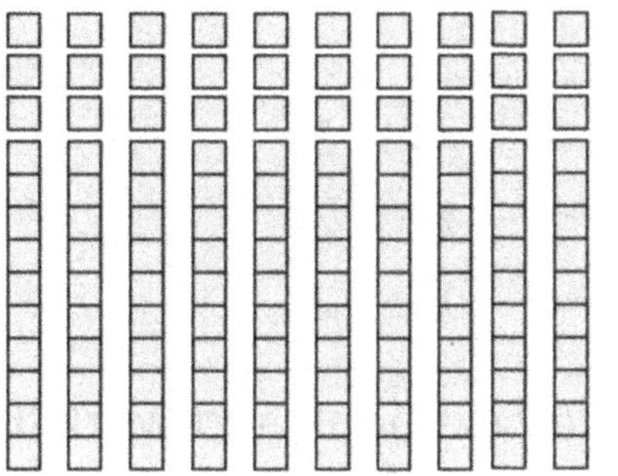

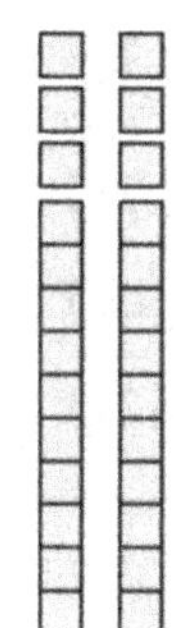

10 lots of 13
130

2 lots of 13
26

B Copy these into your exercise book and work out the answers.

a. 12 groups of 16

b. 11 lots of 15

c. 12 sets of 21

d. 15 lots of 22

e. 14 rows of 24

f. 12 groups of 32

g. 13 rows of 15

h. 13 sets of 21

i. 11 lots of 22

j. 12 groups of 18

k. 13 sets of 13

l. 14 lots of 30

Long Multiplication

$$\begin{array}{r} 34 \\ \times\ 15 \\ \hline 170 \\ 340 \\ \hline 510 \\ \hline \end{array}$$

170 ← 5 × 34

340 ← 10 × 34

510 ← 15 × 34

Copy these sums into your exercise book and work them out as shown in the example at the top of the page.

a. $\begin{array}{r} 23 \\ \times\ 13 \\ \hline \end{array}$ **b.** $\begin{array}{r} 24 \\ \times\ 15 \\ \hline \end{array}$ **c.** $\begin{array}{r} 16 \\ \times\ 14 \\ \hline \end{array}$

d. $\begin{array}{r} 25 \\ \times\ 12 \\ \hline \end{array}$ **e.** $\begin{array}{r} 32 \\ \times\ 16 \\ \hline \end{array}$ **f.** $\begin{array}{r} 26 \\ \times\ 21 \\ \hline \end{array}$

g. $\begin{array}{r} 36 \\ \times\ 24 \\ \hline \end{array}$ **h.** $\begin{array}{r} 52 \\ \times\ 16 \\ \hline \end{array}$ **i.** $\begin{array}{r} 42 \\ \times\ 23 \\ \hline \end{array}$

j. 20 × 16 = **k.** 22× 14 = **l.** 35 × 13 =

m. 15 × 15= **n.** 35× 12 = **o.** 20 × 14 =

Use real objects to check if these pictures are correct.

Why is it important to measure accurately?

Find items that together weigh 1 kilogram. In your exercise book, draw the sets of items.

1. Which items weigh less than 250 g?
2. Which items weigh between 250 g and 500 g?
3. Which items weigh between 500 g and 750 g?
4. Which items weigh more than 750 g?
5. What are some pairs of items that together weigh more than 500 g?

Write the weight of each item in grams.

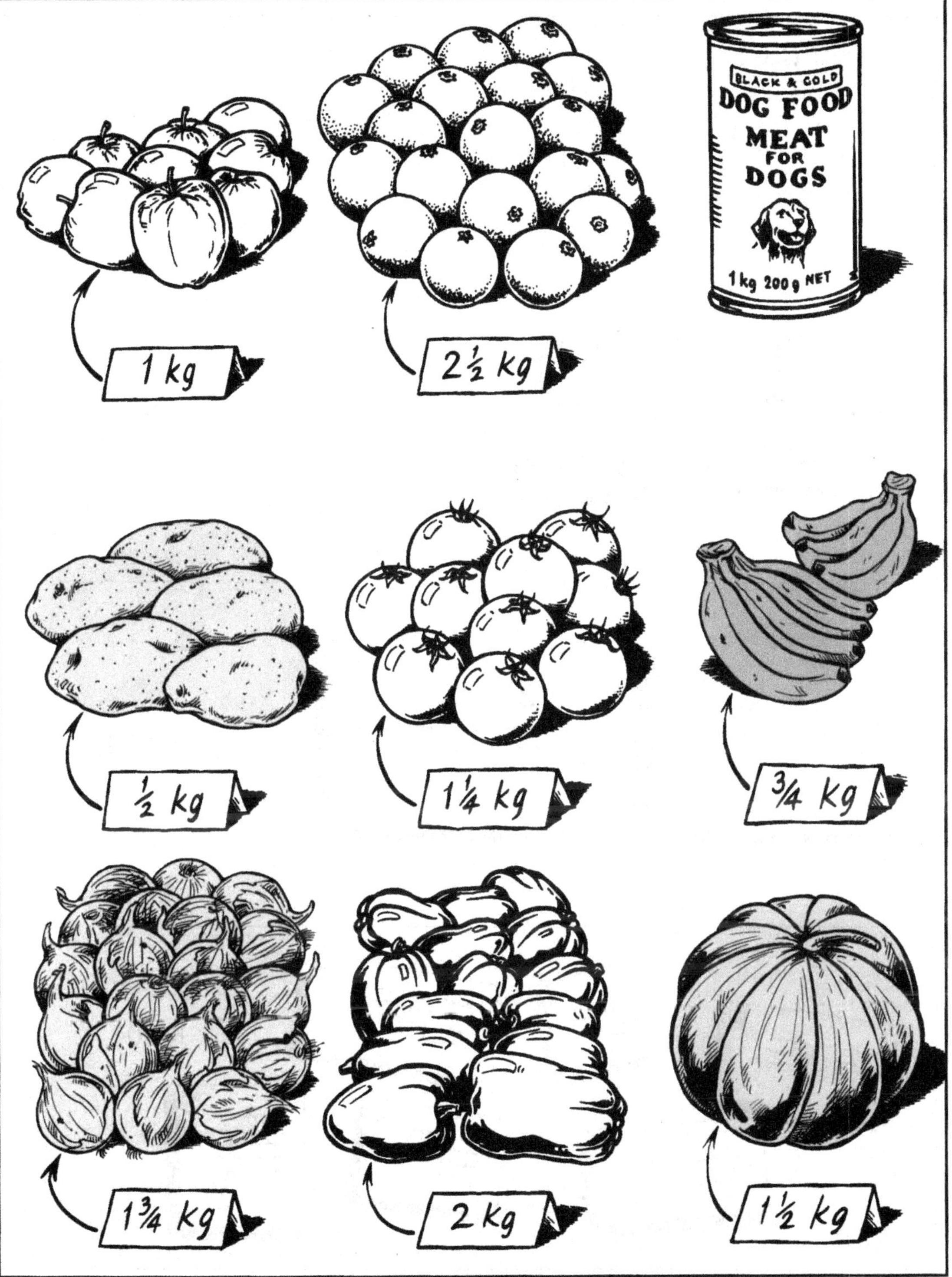

In each Frame, use the total price shown to calculate how much each letter might have weighed.

Weight	Cost
Up to 50 g	25t
51 g to 150 g	58t
151 g to 250 g	K1.00
251 g to 500 g	K2.00

A

K1.00

B

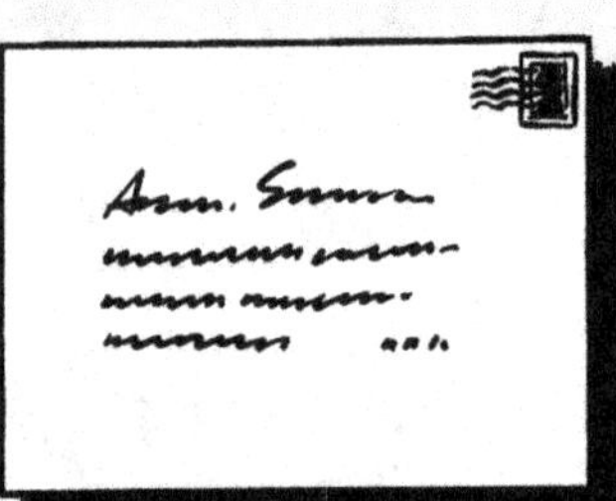

K4.00

C

K6.00

Write the multiplication and division facts for each picture.

Make up a pattern of your own and draw it in your book.

1. There are 35 people and 7 canoes. If each canoe holds the same number of people, how many people will be in each canoe?
2. I need to build a pole which is 18 metres long. I have sticks which are 2 metres long. How many will I need?
3. If three children aged 10, 12, and 14 share 36 pieces of fruit equally, how many pieces will each child get?
4. There are twenty children in a grade 4 class. If they line up in pairs, how many pairs will there be?
5. During a game Tom won 6 marbles, Elsie won 7 and Henry won 8. They decide to share the marbles equally. How many will each child get?
6. I have K24. A two kilogram bag of rice costs K3. How many two kilogram bags could I buy for K24?
7. How many toea will each person get if 27 toea are shared equally between 9 people?

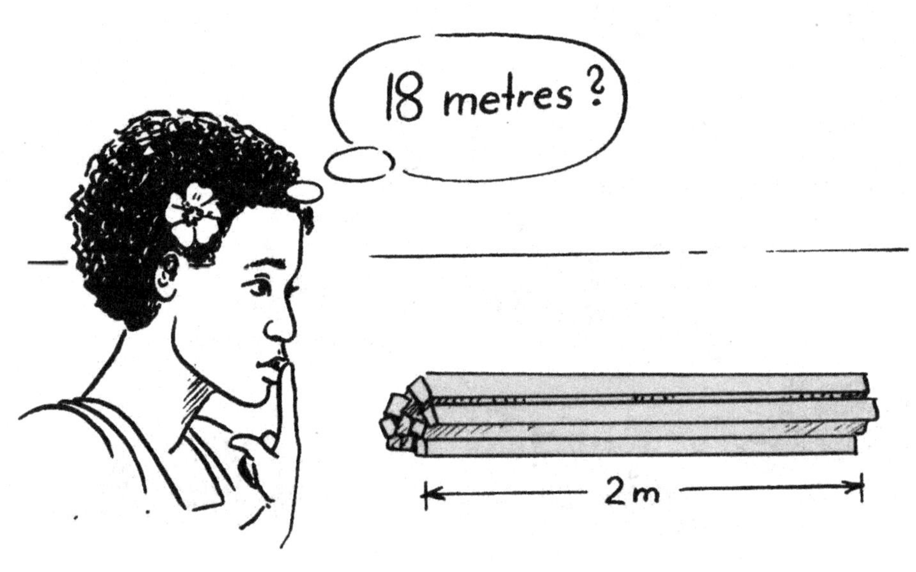

1. A fisherman caught 16 fish. He shared them equally between 3 people. How many did each person get?

2. 27 ÷ 4 =

3. 19 ÷ 5 =

4. Four women collected 17 coconuts. How many bags would they need if they put 3 coconuts in each bag?

5. Two sisters want to share 23 galip nuts. How many will they each get?

6. 35 ÷ 4 =

7. 22 ÷ 6 =

8. 42 ÷ 5 =

9. 25 ÷ 3 =

10. 20 ÷ 3 =

11. 43 ÷ 10 =

12. There are 29 kina. How many people can each get 5 kina?

13. Make up some of your own sentences which have a remainder.

In your exercise book, write these sums and work out the answers.

1. $3\overline{)72}$

2. $6\overline{)67}$

3. $5\overline{)59}$

4. $4\overline{)83}$

5. $76 \div 5 =$

6. $85 \div 6 =$

7. $93 \div 3 =$

8. $61 \div 4 =$

9. $6\overline{)78}$

10. $2\overline{)79}$

11. $5\overline{)67}$

12. $4\overline{)91}$

13. $56 \div 3 =$

14. $27 \div 2 =$

15. $91 \div 5 =$

16. $65 \div 4 =$

17. $6\overline{)52}$

18. $4\overline{)77}$

19. $3\overline{)65}$

20. $5\overline{)89}$

In your exercise book, write these sums and work out the answers.

1. $4\overline{)684}$

2. $3\overline{)672}$

3. $6\overline{)499}$

4. $4\overline{)730}$

5. $164 \div 6 =$

6. $383 \div 5 =$

7. $592 \div 3 =$

8. $607 \div 4 =$

9. $5\overline{)249}$

10. $6\overline{)511}$

11. $409 \div 3 =$

12. $852 \div 4 =$

13. $817 \div 2 =$

14. $419 \div 5 =$

15. $5\overline{)341}$

16. $4\overline{)603}$

17. $3\overline{)754}$

18. $6\overline{)258}$

19. $874 \div 6 =$

20. $328 \div 3 =$

Would this net make the box shown here? Copy the net, cut it out and paste it together. Will your box hold one litre?

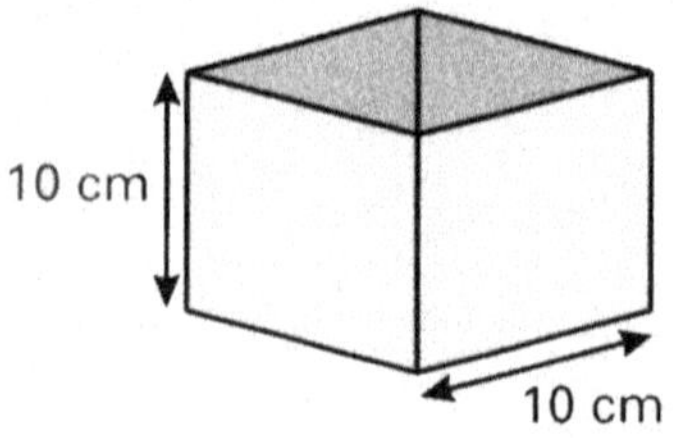

10 cm

10 cm

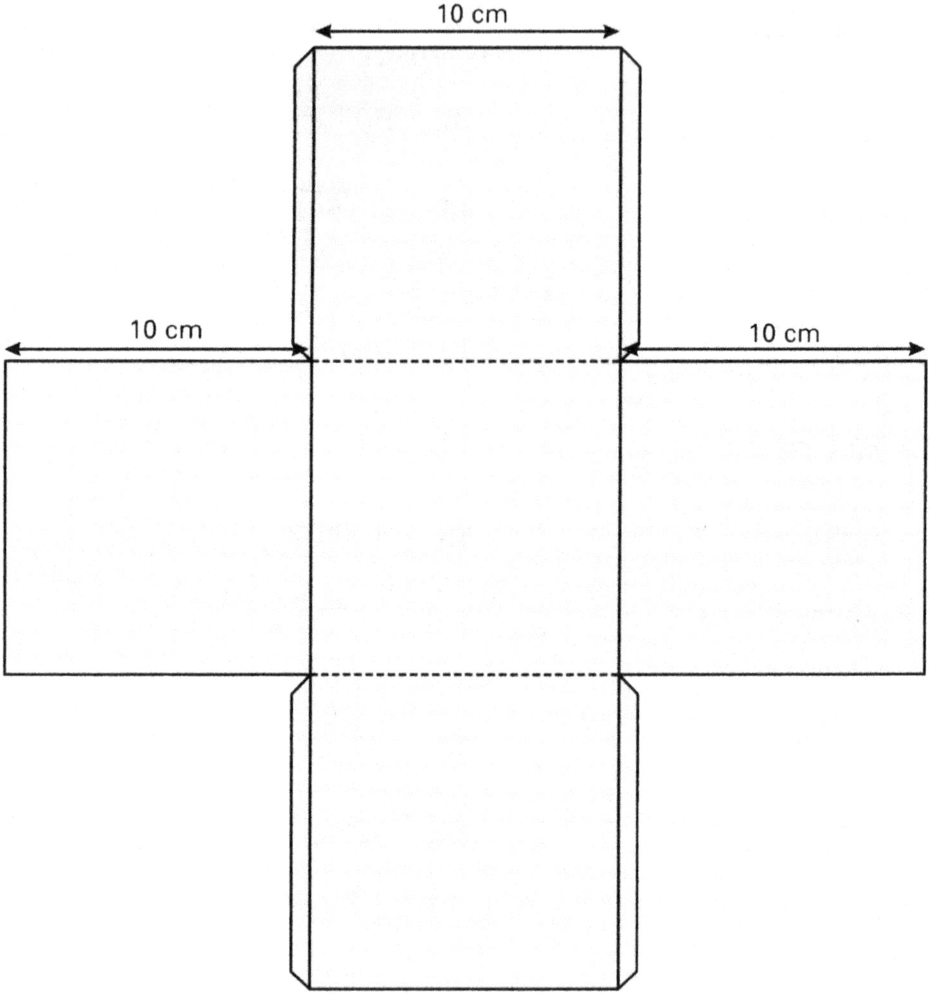

10 cm

10 cm

10 cm

PAULS
NATURAL
LONGLIFE
MILK
UHT PROCESSED
100%
PURE
1 litre
Sno-Wite
BLEACH
1.25 Litres
HEINZ
BIG RED
600 ml
ETA
VEGETABLE
OIL
750 ml
Palmolive
250 ml
ALL PURPOSE
KIKKOMAN
Soy Sauce
1 Litre
SUNCRUSH
RASPBERRY
2 Litres
CAUTION
Sno-Wite
BLEACH
2.5 LITRES
1-Litre
NO ADDED SUGAR
JUST
JUICE
APPLE JUICE
100 % JUICE 100%

How many cubes are needed to make each shape?

In your exercise book, draw these pictures two times as wide and two times as high?

In your exercise book, copy each of these shapes. Write the name of the shape.

circle square triangle hexagon

diamond trapezium rectangle

Different shapes are used to decorate these bilums. What shapes can you see?

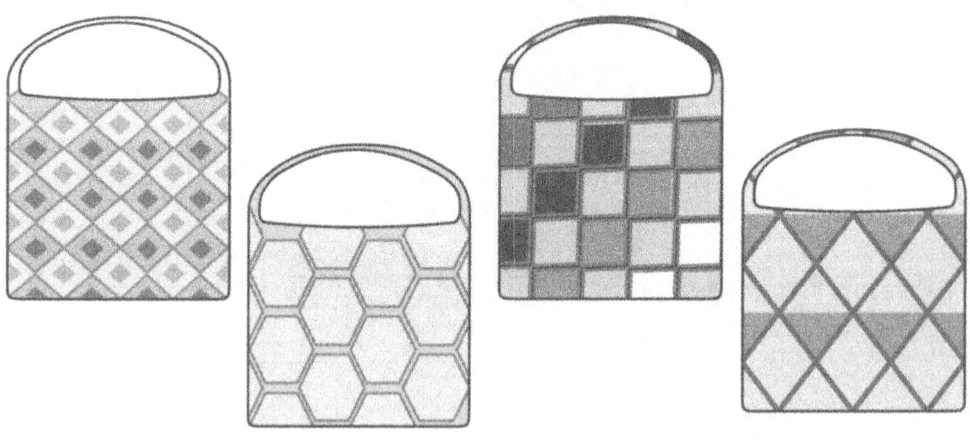

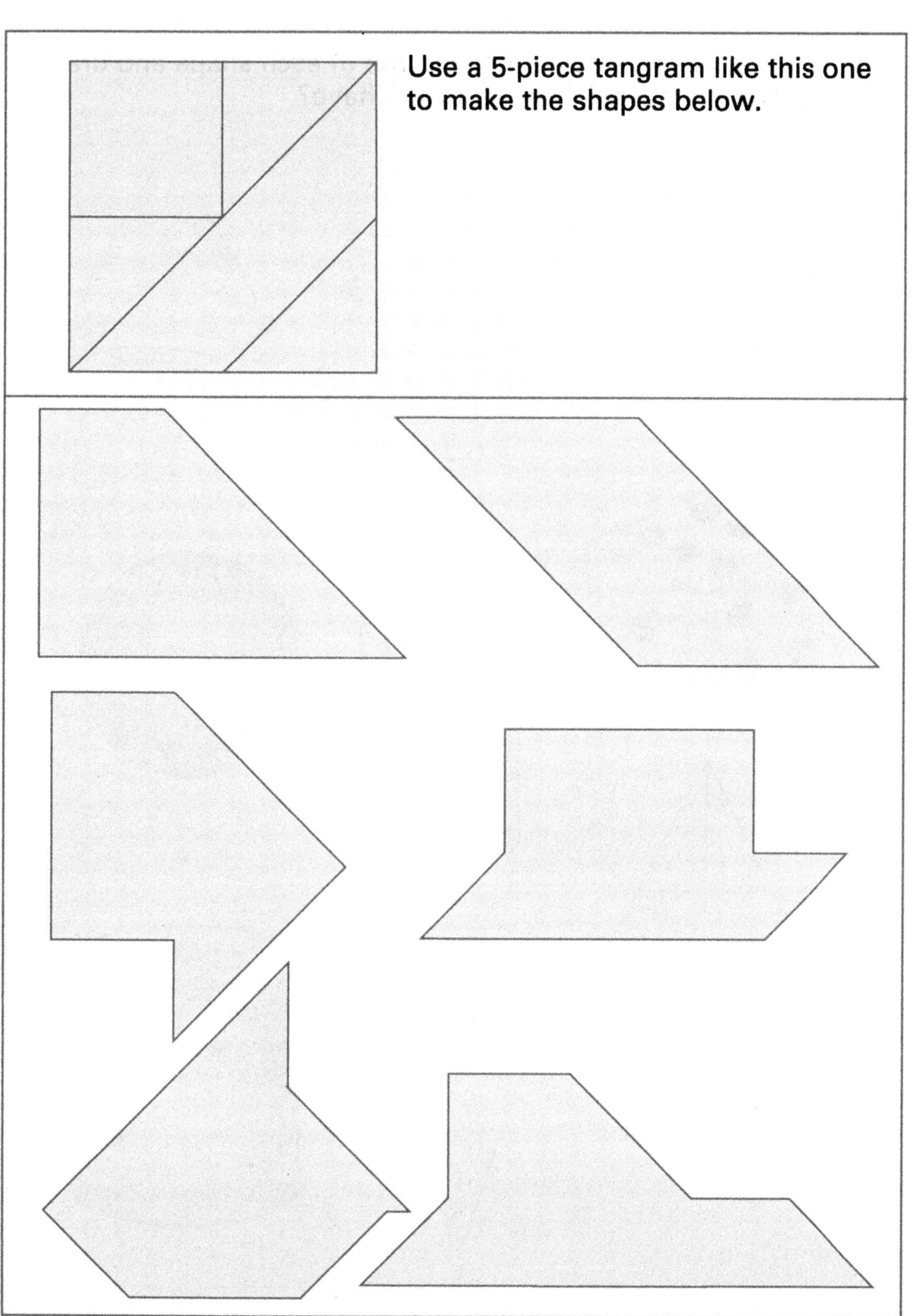
Use a 5-piece tangram like this one
to make the shapes below.

In your exercise book, write the name of each shape and draw it. How many faces does each shape have?

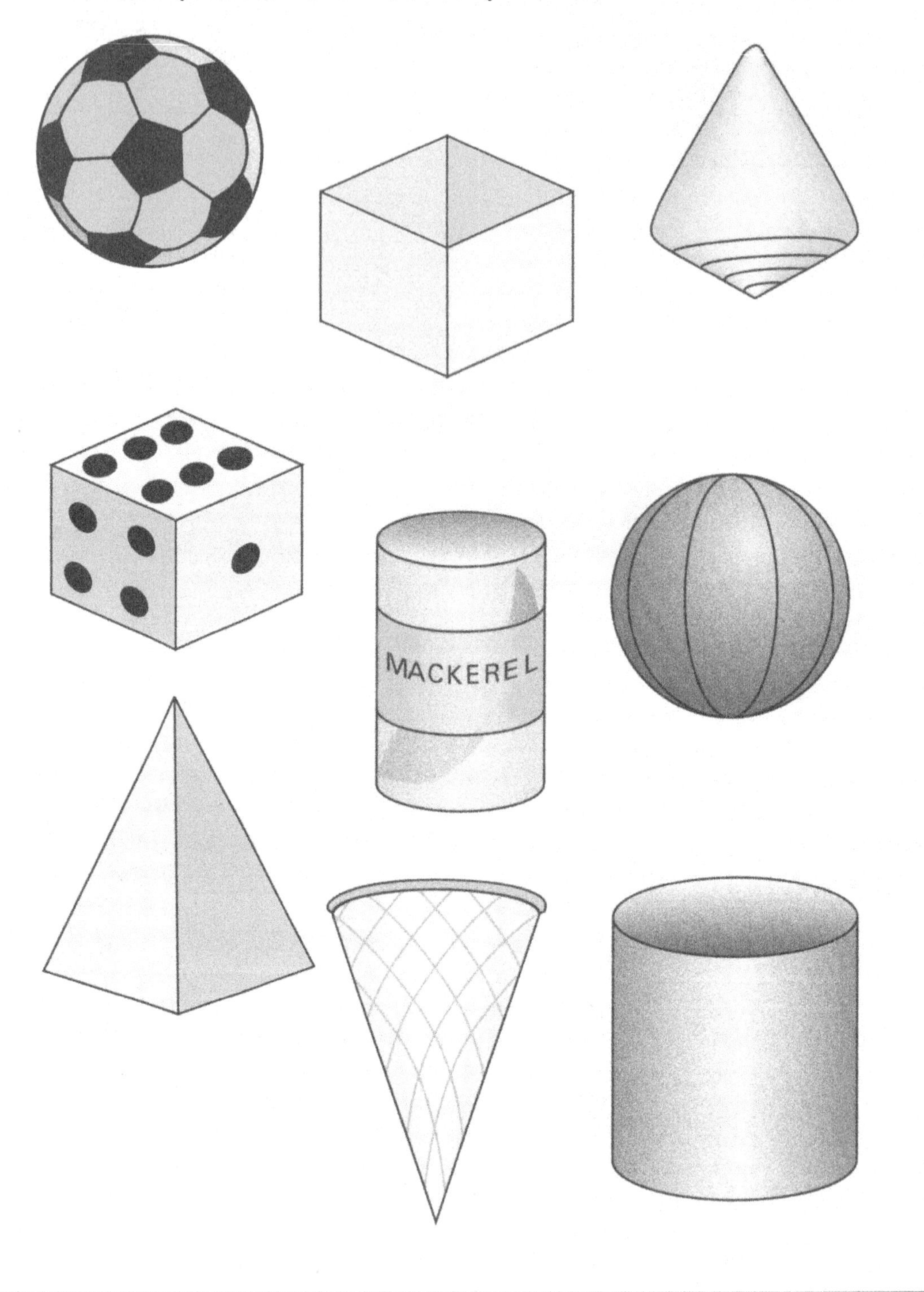

In your exercise book, copy the shapes in Frame A. Next to each shape, write the names of the objects in Frame B which have that shape.

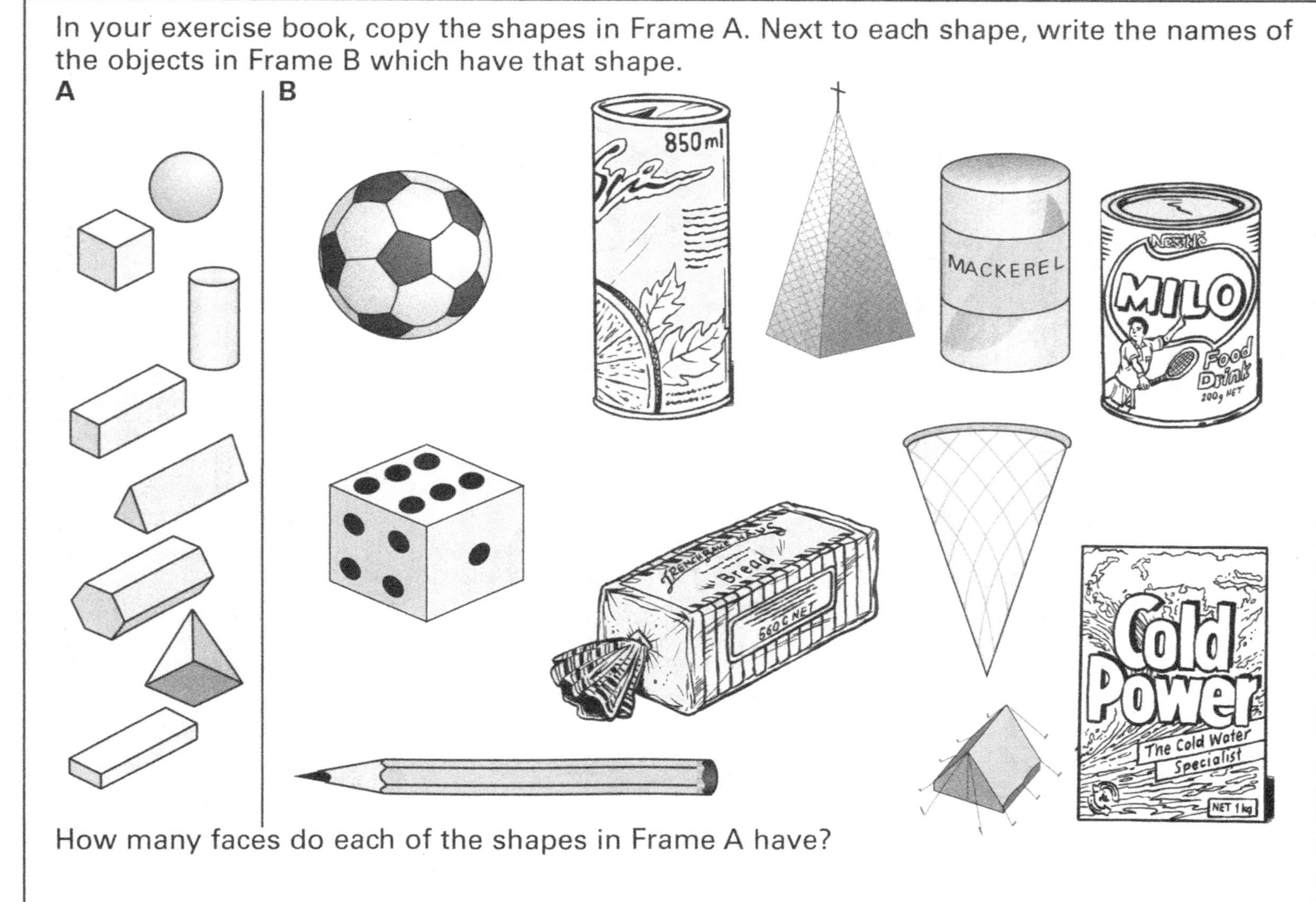

How many faces do each of the shapes in Frame A have?

Draw how each object would look from the side.

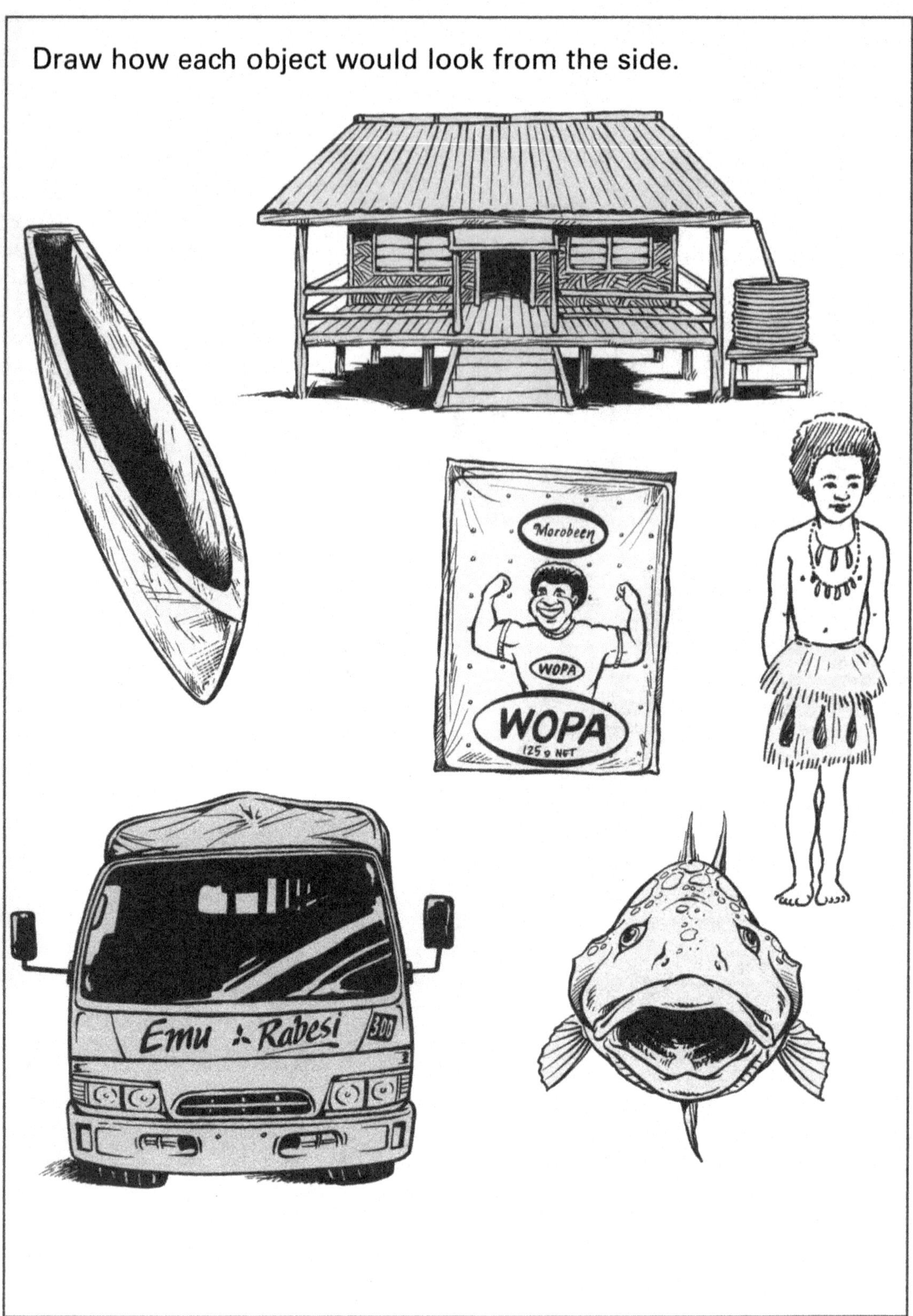

Use the ruler to estimate the length of each object. Check with your own ruler.

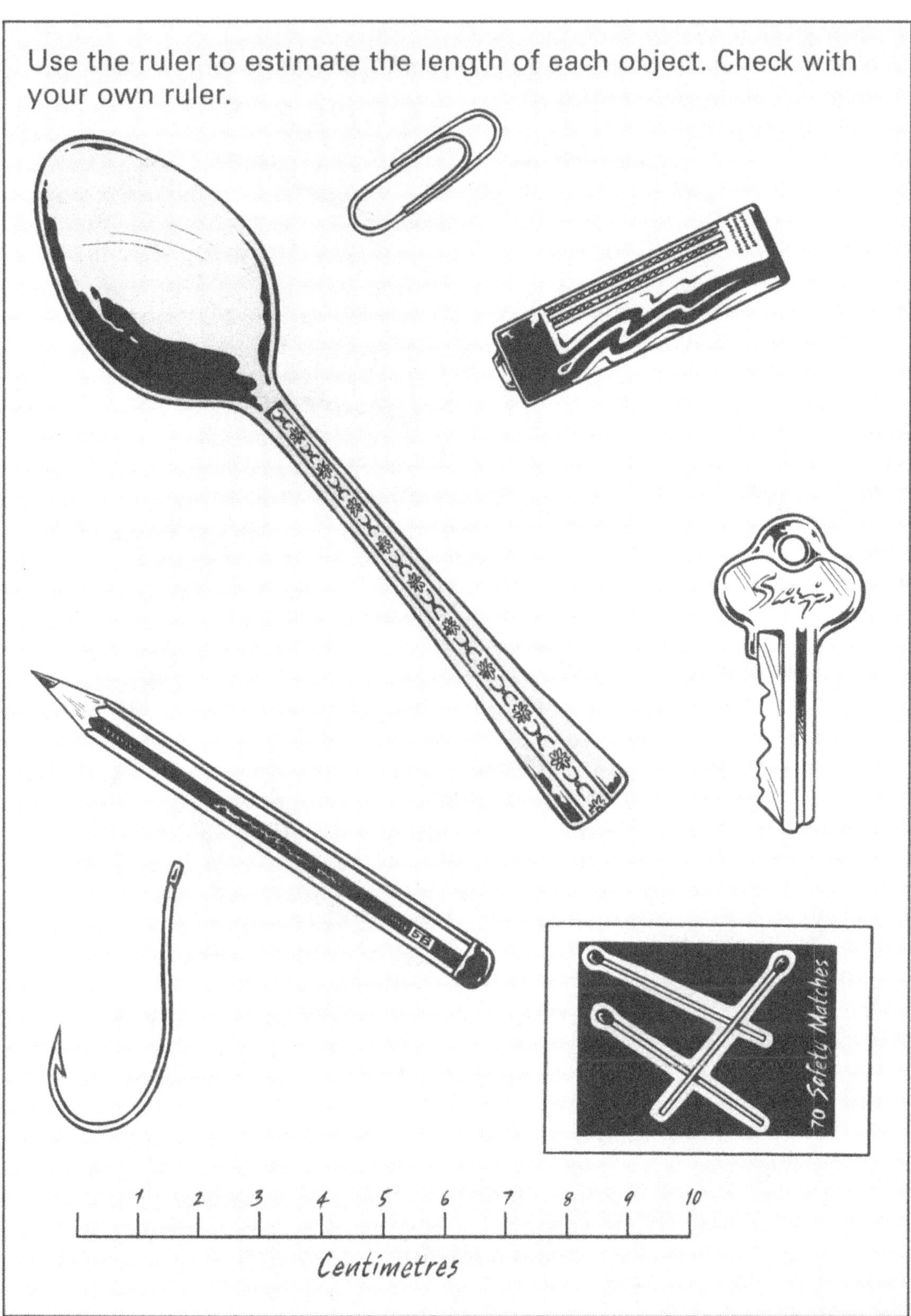

Use the ruler and some string to work out the length of each piece of string on the page.

1 2 3 4 5 6 7 8 9 10 11 12

Centimetres

Find the perimeter of each shape.

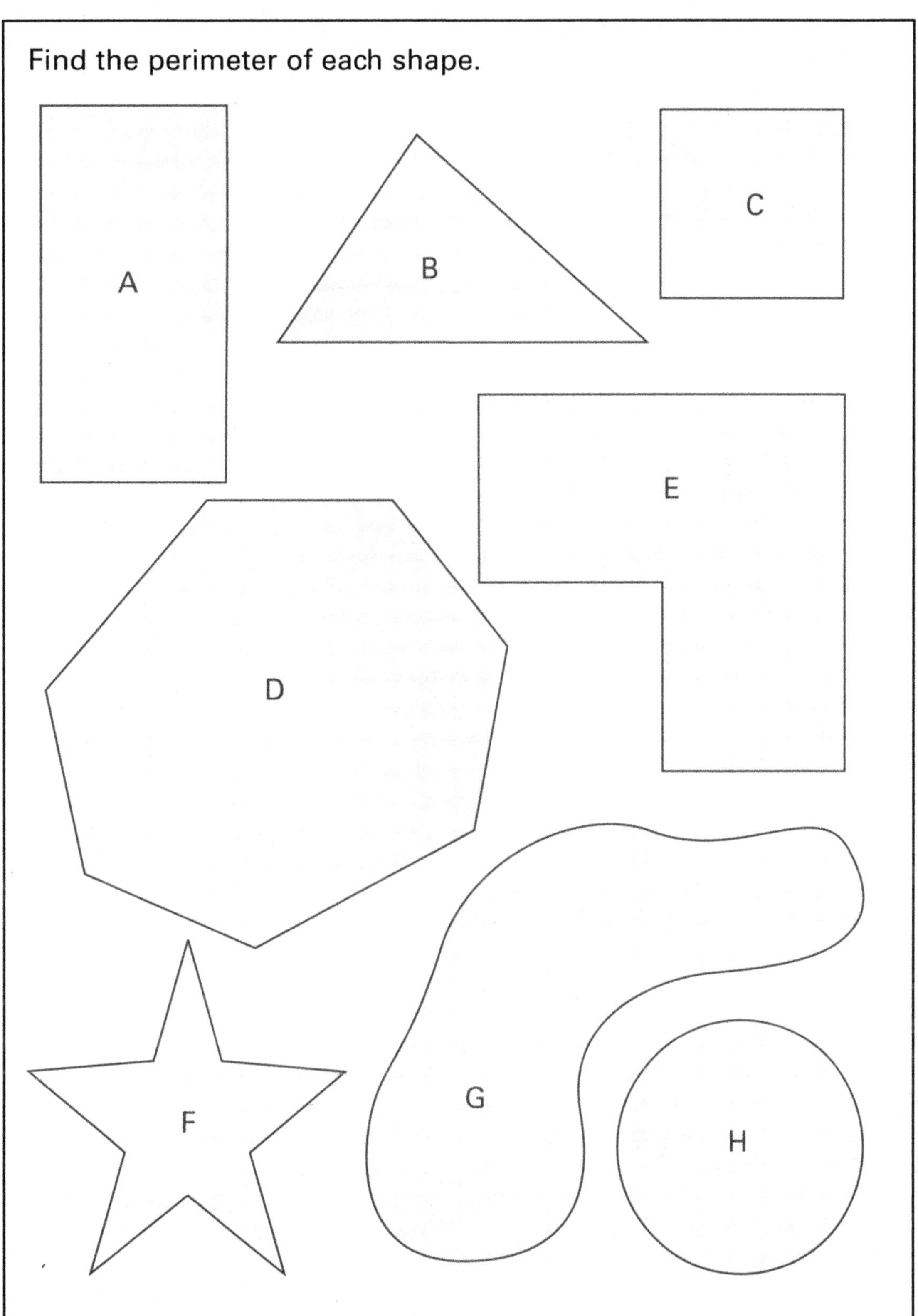

Measure each of these distances on your own body and write the answers in your exercise book.

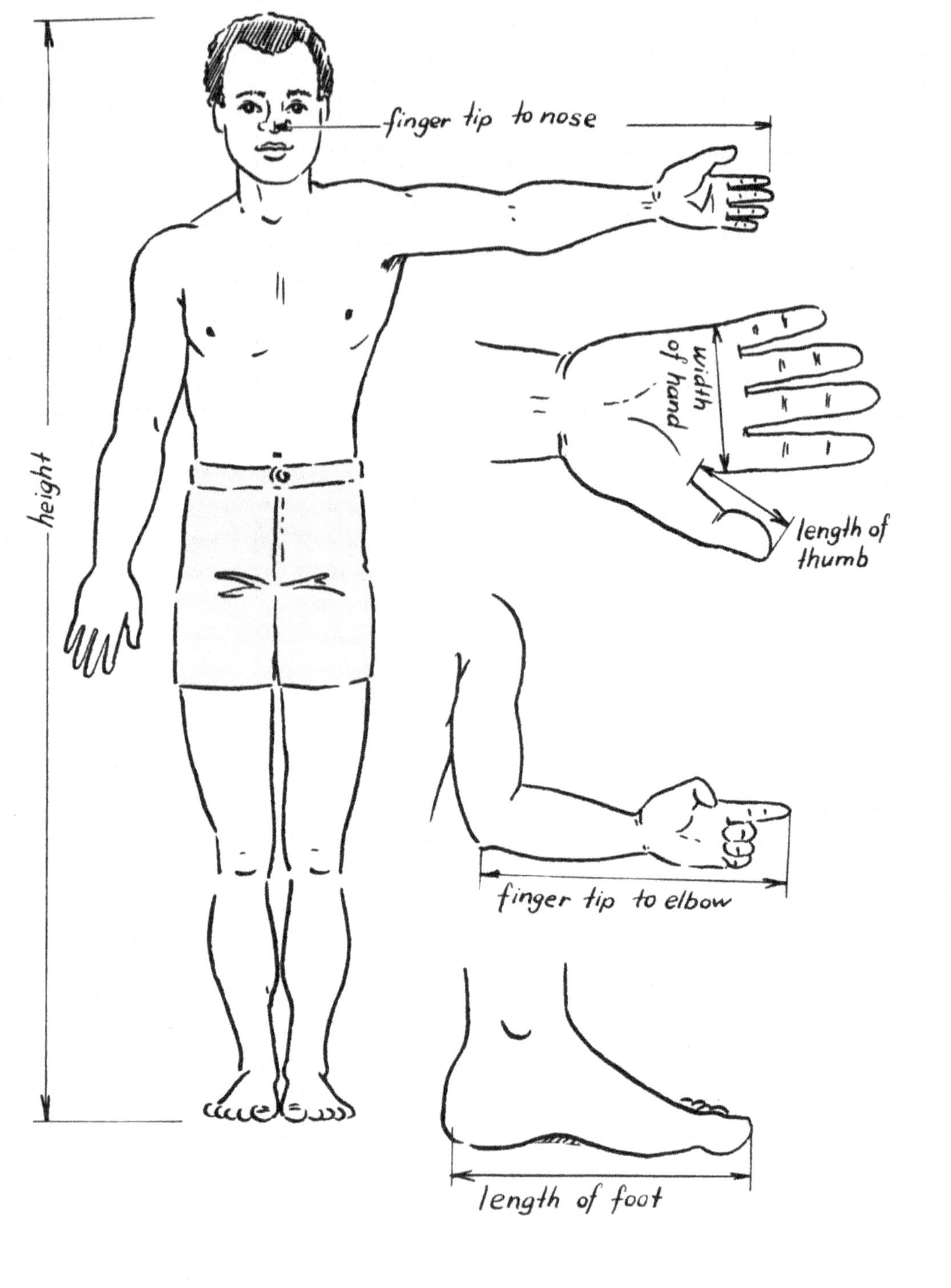

Estimate the length of each line and then measure each one.

Write each record in centimetres.

South Pacific Games Record is 7.49 m

South Pacific Games Record is 2.05 m

South Pacific Games Record is 15.66 m

South Pacific Games Record is 51.32 m

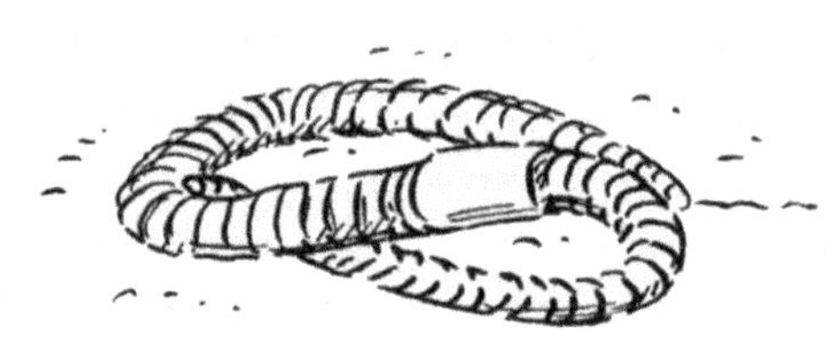

World's largest worm is 6.7 m

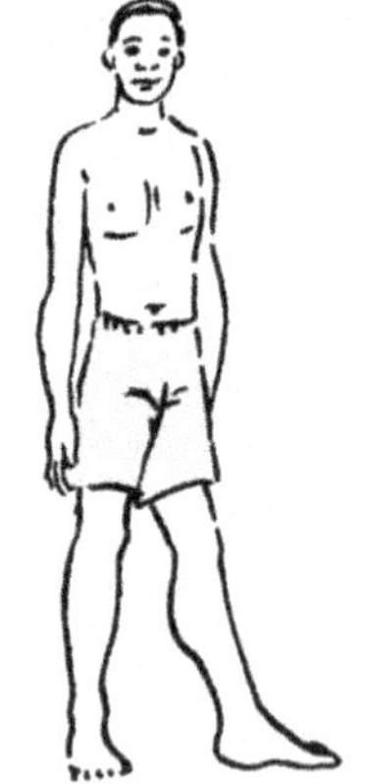

World's tallest man is 2.72 m